DIONES REIS

Didactics & CHEMICAL LINKS SEQUENCE

DIONES REIS

Didactics & CHEMICAL LINKS SEQUENCE

Challenges and Possibilities

ScienciaScripts

THANKS

To God for the opportunity and the wisdom he had given me, for without him I would not have succeeded.

To my parents, Josefa Maria do Reis and Gilvan Bento dos Reis for believing and supporting me. To my sisters Katiana Maria dos Reis and Denise Bento dos Reis.

To my son Samuel Reis Costa with whom I was blessed by God, and he makes me immensely happy.

To my fiancée, Jussara Faustino dos Santos for her strength, help and constant encouragement, even when I was absent, for being a person present in my life, even when I failed her. For the opportunity of a new beginning. And for accepting me with my qualities and defects.

To all the members of the "Benedict of Kings", "Faustino dos Santos" and "Matias e Cardoso" families for their affection and love.

To the students who participated in the research, for their commitment to the classes and activities carried out.

To Professor Pedro Lucio Barboza, for the advice, encouragement and corrections that were primordial for the conclusion of this work, for having accepted me as his mentor and for having believed in me, more than guiding a friend.

To the examining board Prof. Eduardo Gomes Onofre for his words of encouragement during my career in the course and to Prof. Luiz Havelange Soares for his participation and contribution in the board.

To the teachers of the course that I had the opportunity to meet along the way.

To her friends Emanuelle Maria Cabral, Wagner Ananias, Cláudia Correia. Joan Darc and Julius Cesar for friendship, support, affection.

To the sports friends for the moments of joy and relaxation, helping in this way to forget the problems and stress to focus on the conclusion of the course.

To the colleagues of the course, who provided wonderful days in my passage through this institution.

SUMMARY

Chemistry is a very beautiful and fascinating science, but unfortunately little understood by students. It still has great resistance, especially in high school students, even in primary schools when there is the first contact(s) with chemistry. This research is therefore of a qualitative nature, with the aim of analysing the performance of students in the first year of high school, following the application of a Teaching Sequence developed in 6 (six) classes (stages), in the teaching of Chemistry Links. We used a pre-questionnaire composed of closed questions, which contained information concerning the object of the research in order to have a previous verification of the students' knowledge(s). We used a field diary, where some students' and teachers' statements were recorded. We requested a textual production on the subject. And finally, we apply a final written activity based on the molecular models elaborated by the students representing the types of Chemical Bonds. In the subjective questions (textual production and final activity) we used discourse analysis and the objective questions (questionnaire) were placed by means of graphs and charts, and later discussed. We also used discourse analysis for the statements of students and teachers recorded during the application of the Didactic Sequence. Also, in the final activity we built a board where we made the synthesis of the students' answers. In addition, the images recorded of the students during the activities of the Didactic Sequence were discussed. The results show that the students liked the didactic proposal, favouring the teaching-learning process. We can conclude that the objectives were achieved, as the students were able to critically position themselves and to conceptualize some processes of the content of Chemical Link. We also realised that the Didactic Sequence is an important resource in the teaching-learning process and can be applied in the classes as a "motivational and helping tool" for the students, once it is worked in an active perspective, promoting the dialogue (the discourse of the theme) of the students with each other, leading them initially about the content approached, having as intention to awaken their autonomous capacity of questioning the world around them.

KEY WORDS: Chemistry Teaching; Chemical Bonding; Didactic Sequence.

SUMMARY

ACKNOWLEDGEMENTS 2

SUMMARY 3

ABSTRACT 4

LIST OF ACRONYMS 5

SUMMARY 6

INTRODUCTION 7

1. ON THE PROCESS OF TEACHING AND LEARNING CHEMISTRY 8

2. TRAINING OF CHEMISTRY TEACHERS 22

3. THE TEACHING SEQUENCES 29

4. METHODOLOGICAL ASPECTS 35

5. RESULTS AND DISCUSSION 39

6. CONCLUSIONS 61

REFERENCES 63

APPENDIX I 68

APPENDIX II 69

APPENDIX III 71

APPENDIX IV 73

INTRODUCTION

For a quality teaching we have to consider some aspects that are very important, and are linked to learning, such as: teacher training, time of experience, educational space, time of assimilation of the student on the content taught (each student has a time and processes the information at different speeds).

I don't know what I want. I don't know what I like. These, and so many others, are dilemmas that many students face, or face these enquiries, and cannot find something, or awaken the interest to remain in the search for completion and continuity in school life. This explains, in part, why there is an evasion by many high school students, who are unable to complete it because of a lack of perspective.

Many students find it difficult to learn chemistry at different levels of education because they do not understand the meaning or validity of what they are studying.

Chemistry is undoubtedly fascinating in particular, but little understood. It still shows a lot of resistance, mainly by students in high school. It is still evident in primary schools, when the "first contact" with chemistry occurs. It is a fact that in primary schools chemistry is linked to science teaching, so there is no dissociation with the other sciences. In other words, there is no dissociation of chemical concepts. This is accentuated by the absence of any connection between what the teacher tries to teach and the daily life of the student, when they begin their studies in the first year of high school, at a time when there is a "break" between the disciplines, i.e., at this level, the sciences come together into disciplines that are taught separately, when in nature and in the daily life of the student they coexist simultaneously.

This particular view (mine) of the way many students view chemistry, and the limitations on their learning of it, was experienced, and observed during my graduation from the State University of Paraíba when I participated, for a period of 2 (two) years, as a scholarship student, in the Teaching Initiation Program, which was made available to the undergraduate courses. This experience in public schools gave me the opportunity to understand the difficulty some students had in understanding the content of chemical links.

From this problematization I have produced the course conclusion work (Monograph), in which we make available as a teaching proposal a Didactic Sequence for future research.

In this sense, this research has a qualitative nature and aims to analyze the performance of students in the 1st year of high school, after applying a Didactic Sequence based on the use of molecular models in the teaching of Chemical Bonds.

We also seek to understand the meaning and validity of using molecular models with active function in the classroom.

This dissertation is structured in six chapters.

In chapter 1: We will discuss the process of teaching and learning chemistry. This is divided into topics: 1.1 Teaching Science through Models; 1.2 Teaching Chemistry through Models; 1.3 Epistemological Obstacles in Teaching Chemistry in the Use of Models by Students and Teachers; 1.4 Teaching Chemistry Links. It is subdivided in the topic: 1.4.1 Teaching Chemistry through models: possibilities and limitations.

In Chapter 2: We will deal with the training of chemistry teachers. This is divided into the topic: 2.1 What chemistry teacher training do we want?

In Chapter 3: We will review Teaching Sequences. Where we will do a bibliographic review bringing concepts and some authors who work with this teaching proposal.

In Chapter 4: We will deal with the methodological aspects. We present the methodological references that have guided our bibliographical review study. This is divided into the following topics: 4.1 Qualitative research; 4.2 Characterization of the interlocutors of the research; 4.3 School characterization; and 4.4 Data collection.

In Chapter 5: Results and Discussion, presents an analysis of the productions made by the students and its treated in class.

In Chapter 6: We will present our concluding remarks.

1. ON THE PROCESS OF TEACHING AND LEARNING CHEMISTRY

With the Natural Sciences having in common research on nature and the development of its technologies, it is with this vision that the school should share and articulate the languages that make up each scientific culture, establishing measurements capable of producing knowledge, in the dynamic interrelationship of diverse daily and scientific concepts, including the cultural universe of each Science.

Like other natural sciences, chemistry participates in scientific and technological development with important contributions, having economic, social and political scope. In this sense, society and its citizens interact with chemistry by various means.

The old emphasis of "traditionalist education" proposed that through the transmission of information, the student, by memorizing it, acquires solid knowledge that can help him to understand the world around him. However, it is clear from the NCPs (1999) that the proposal put forward for the teaching of chemistry is opposed to the memorisation of information, names, formulas and knowledge as fragments disconnected from the reality of the students. On the contrary, it advocates that the student recognizes and understands, in an integrated way and significativa, the chemical transformations that occur in natural and technological processes in different contexts. Thus, the learning of chemistry in high school should enable the understanding of both the chemical processes themselves and the construction of knowledge científico in relation to technological applications and their environmental, social, political and economic implications. In this way, students can judge on a sound basis information coming from cultural tradition, the media, and the school itself and make decisions autonomously, as individuals and as citizens.

According to the NCPs+ (2002), the teaching of chemistry should be understood from the following perspective:

> Chemistry can be an instrument of human formation that broadens cultural horizons and autonomy in the exercise of citizenship, if chemical knowledge is promoted as one of the means of interpreting the world and intervening in reality, if it is presented as science, with its own concepts, methods and languages, and as a historical construction, related to technological development and the many aspects of life in society (BRAZIL, 2002, p.87).

From the NCPs+ (2002), we can observe that historically, chemical knowledge was based on empirical studies on the properties of materials and substances and chemical transformations. In this perspective, the learning of chemistry, provides the opportunity to develop skills and abilities, emphasizing real problematic situations in a critical way. Therefore, it provides the student with skills such as interpreting and analysing data, arguing, drawing conclusions, evaluating and making decisions.

Thus, in the organization of the subjects that the school must address, that is, the selection of what to teach, must be guided by the choice of contents and relevant themes that allow the understanding of the natural, social, political and economic world. In order to achieve this objective, the way in which these themes and contents are dealt with is determinant and they must include the development of procedures, attitudes and values. From the moment knowledge is built with this scope, and in an integrated way with other sciences and fields of knowledge, within real contexts and considering the formation and interests of each student, it will be favouring the development of the different competences proposed in the PCNEM. In this way, it can be stated that the competences and skills to be developed in the area of Nature Sciences, in particular Chemistry and its Technologies, concern the domains of representation and communication, research and understanding and socio-cultural contextualisation.

The following table summarizes the role that schools, according to the PCNEM (1999), need to develop around skills and abilities in chemistry teaching. Let's see:

Table 1 - Skills and Skills in Chemistry Teaching.

COMPETENCES AND SKILLS

1) Representation and communication

- Describe the chemical transformations in discursive languages.
- Understand the codes and symbols of today's chemistry.
- Translating discursive language into symbolic language of chemistry and vice versa. Using the symbolic representation of chemical transformations and recognizing their modifications over time.
- Translating discursive language into other languages used in chemistry: graphics, tables and mathematical relations.
- Identify sources of information and ways to obtain information relevant to the knowledge of chemistry (book, computer, newspapers, manuals etc.).

2) Research and understanding

- Understand and use chemical concepts within a macroscopic (logical-empirical) view.
- Understand the chemical facts within a macroscopic (logical-formal) vision.
- Understand quantitative data, estimation and measures, understand proportional relationships present in chemistry (proportional reasoning).
- Recognise trends and relationships from experimental or other data (classification, seriation and correspondence in chemistry).
- Select and use scientific ideas and procedures (laws, theories, models) to solve qualitative and quantitative problems in chemistry, identifying and monitoring the relevant variables.
- Recognising or proposing research into a problem related to chemistry by selecting relevant experimental procedures.
- Develop hypothetical logical connections that allow predictions about chemical transformations.

3) Sociocultural contextualisation

- Recognize relevant chemical aspects in the individual and collective interaction of human beings with the environment.
- Recognize the role of chemistry in the productive, industrial and rural system.
- Recognise the links between the scientific and technological development of chemistry and socio-political-cultural aspects.

Source: Ministry of Education (MEC), Secretariat of Media and Technological Education (Semtec), (BRAZIL, 1999).

In this sense, the skills and competencies enunciated by the PCNEM (1999), proposed for the teaching of chemistry should be associated with the contents to be developed, being an inseparable part of these contents, and should be effective from the different themes proposed for the study of chemistry, at levels of comprehensiveness compatible with the subject addressed and with the level of cognitive development of students.

However, the different educational and social realities of students differ in their perceptions of chemical knowledge, thus generating the need for pedagogical actions that can "empower" the teaching-learning process.

A serious problem in the teaching and learning process of chemistry, as a subject, refers to the transposition of the contents worked by the teacher and the lack of assimilation of the contents by the students. Difficulties that are evidenced in the student's evaluation. In other words, is the student's learning or the teacher's teaching method being evaluated? Are these difficulties of assimilation arising from the way the teacher approaches the content? Or even, no matter how elaborate the teaching methodology, as well as the motivation, are such "barriers" internal? These are discussions pertinent to the "conflicting relationship", between teacher-student, as well as the relationship between the teaching-learning process, especially of chemistry.

From this perspective we have to consider the *"multidimensions"* of the student. The teacher should have the sensitivity to understand that each student has his or her own particular way of thinking and acting; and these may be influenced by internal factors (such as emotional, psychological and physical) and factors external to the student that may also influence their performance (such as economic, social, religious and political). The role of the teacher, in its wider scope, soon gains the role of the educator. An effective initial formation is essential for this.

Another observed situation is that teachers are not always prepared to act in an interdisciplinary way, integrating the chemical contents with the reality of the students. This occurs due to the lack of qualified training of the teachers who take over the chemistry classes, or even due to the occurrence of teachers with another academic background.

Since in science teaching, particularly chemistry, students at many times do not understand or are unable to associate the content addressed with their daily lives, causing it to lose interest, this reveals the lack of approximation of the content to the reality of the student, or even the lack of understanding of the environment around them. The need to prioritise the teaching-learning process in chemistry in a contextualized way, in accordance with what the PCNEM recommends, should be observed.

This lack of understanding by students is due in part to the distorted view that many hold of chemistry as the villain among the natural sciences, not only shows a prejudice rooted in society, and especially among students, but also points to a deficiency in its teaching in educational institutions.

Santos *et al* (2015), discusses that:

> The public image of chemistry after five scientific revolutions, Chamizo [12] concludes that in addition to being useful, chemistry is also considered an impure and dirty science. Another category that deserves a lot of emphasis is that chemistry is a polluting science, being considered responsible for situations that harm man and the environment. Thus chemistry still faces a very hostile environment in today's society (SANTOS *et al*, 2015, pp. 49-57).

The present study was a research carried out through the analysis of the contents of the online articles of the journal Ciência Hoje, category of Chemistry, published in the years 2011 and 2012, where the main objective was to identify the public image of Chemistry in these articles from the categories obtained in the initial research. The analyses of Santos *et al*, showed the positive image that many have about Chemistry, but also highlighted the bad vision that exists about it. This distorted view contributes to the mystification that everything about chemistry is bad. Many students come to school with the conception that Chemistry does not need to be learned, since they cannot visualize its application in their daily lives. However, as science, we must teach its industrial functionality, but also its importance in the daily life of the student, permeating the construction of knowledge through the rupture between "popular knowledge" and "academic knowledge", thus, it would be opportune for a more effective learning.

We should not see the teaching-learning process as a uniform aspect, but observe it according to the relationship between the teaching process and the learning process, applied and analysed, enunciating the educational path in the production of knowledge. Thus also the teacher-student relationship cannot be evaluated simply in its singularity, i.e. in its unitary value, but proceeding with the teacher-student relationship(s) and the pupil-student relationship(s). The classroom becomes comprehensive when we understand that in the school universe there is cultural diversity and where the individual (teacher and/or student) must be seen in their emotional, psychological and physical plurality. In this way the educational process cannot be limited to traditional pedagogical practices hoping that it will have the same effect on the generation of today's students, since we live in a time that is so "attractive" as to distract them by providing disinterest in the great majority.

In general, students have a very hard time understanding, so they don't understand chemical concepts. This creates an obstacle(s), causing students to have a bad view of Chemistry, because the student who does not understand it, cannot dazzle how beautiful it is as a science.

As educators and researchers in education, we must be aware that it is "possible and necessary" to improve the quality of chemistry teaching. In Brazil we have a great lack of professionals in this area, which in recent years is one of the fastest growing in research. Today many, if not all universities (public and private) have approached and invested in research in the teaching of chemistry, and this is good for the area, because it is now recognized, to have more importance, favoring a new vision of chemistry teaching in Brazil.

In the teaching of chemistry, it is not enough just to have knowledge of this Science, it is important to have knowledge of Didactics, Pedagogy Theories, Psychology, among others, because these are subjects that will present the possibility of teaching with better quality. To reach this level of teaching it is necessary to know how the individual learns, what motivates him/her to learn a certain subject, what arouses the interest in learning chemistry.

The teaching-learning process within the school is not only linked to the technical concepts that we observe in the approach of various contents in Mathematics, Physics, Biology or even Chemistry, it is also linked to the whole social concept that the student and the teacher live. And this influence within the classroom both in the teaching by the teacher and in the learning by the student.

We must demystify in the student's conception that the sciences (Chemistry, Physics, Mathematics, Biology) are irrefutable truths, and therefore it is not a mechanized knowledge. The questioning is necessary, that the students are involved and realize that the understanding (even if initially) on the subject approached occurs with the question (enquiry) accompanied by the answer (opinion) of the same, even if erroneous, but the "self-criticity" is important.

In this sense, such behaviour must exist for the various contents of chemistry, such as Chemical Links.

1.1 Teaching Science through Models

There are different meanings for the word Model. In the educational literature we find different contributions from different authors in order to help meet the concept of models. From a brief discussion, that is, without seeking a tiring discussion, we will present the considerations of some authors about what models represent as put forward by Lima and Nuñez (2004):

Table 2 - Model concept according to some authors.

Castro (1992)	The models represent a particular image of an aspect of reality and by definition would be incomplete, in relation to the system it is intended to represent (referent or object system) which is normally a complete system.
Pozo and Crespo (1998)	Models are a representative process that makes use of images, analogies and metaphors, to help the subject (student or scientist) to visualize and understand the referent, which may present itself as difficult to understand, complex and abstract, or in some perceptibly inaccessible scale.
Galagovsky and Adúriz-Bravo (2001)	Models are considered to be theoretical tools for representing the world, helping to explain, predict and transform it.

Source: Taken from the Book, Fundamentals of Teaching Learning Natural Sciences and Mathematics: The New High School (2004).

We can consider the models, based on the authors' reflections, as a representation of reality, which makes it possible, in the scientific area, to discover and analyze (study) new interactions and aspects of the object of study, being mutable and limited representations, in relation to the complexity of the phenomena they seek to represent.

In the study of science, in different fields (chemistry, physics, biology, mathematics, etc.), there is the constant use of representations in order to understand and explain the phenomena involving abstract concepts. Thus, since science is changeable, these representations appear with a provisional character, being called a model. From these, theories and laws are proposed, based on the model initially developed. In the case of chemistry, through models scientists formulate questions about the world. As well as through the use of Models, the teachers provide an opportunity for the interaction of the student's vision with the content approached, making the class more attractive from the moment it manages to arouse the student's interest. In this way we see the illustrative models as a good "tool" to help the teacher in the teaching-learning process.

1.2 Teaching Chemistry through Models

According to Morais (2007), some researchers have pointed out different approaches to concepts in the teaching of chemistry, such as "reformulation" of teaching strategies based on the integration of laboratory activities in classes, use of concrete models and use of technologies as learning tools.

The author highlights that several studies have indicated good learning results when using concrete molecular models as a way of visualizing the particle model and the associated chemical transformations. This type of visualization is pointed out as one of the most used nowadays, because it simplifies, illustrates and allows the exploration of the structure and the associated chemical process. However, these models are rigid, and generally in limited quantity, which restricts their use to the representation of small molecules. The teacher's intention in using models is to facilitate the student's understanding, and consequently provide meaningful learning.

> For all practical purposes, the acquisition of knowledge in teaching depends on verbal learning and other forms of symbolic learning. In fact, it is largely due to language and symbolisation that most complex forms of cognitive functioning become possible (AUSUBEL, 1968, p. 79).

According to Ausubel (2003) in Meaningful Learning the ideas expressed symbolically interact in a substantive and non-arbitrary way with what the learner already knows, that is, it means that the interaction is not with any previous idea, but with some specifically relevant knowledge already existing in the cognitive structure of the learner relevant to the new learning, which can be, for example, an already significant symbol, a concept, a proposition, a mental model, an image, called subsumption or idea-anchor.

The term subsumption is the specific knowledge, existing in the subject's structure of knowledge, which allows to give meaning to a new knowledge that is presented to him or discovered by him. Both by reception and by discovery, the assignment of meanings to new knowledge depends on the existence of specifically relevant previous knowledge and the interaction with it. In this process, the new knowledge acquires meaning for the subject and the previous knowledge acquires new meanings or greater cognitive stability.

According to Moura and Morretti (2003) the most impacting aspect that influences learning is what the student already knows, i.e. their previous knowledge. Therefore, this should be investigated and the teaching should start from this data. From this point of view, we understand that for significant learning to occur, it is necessary that the teacher uses a good teaching methodology, which can be easily understood by the student, and makes use of a potentially significant material, in this particular study, the Didactic Sequence by models (Or,

previous daily culture, since it is not possible to incorporate new knowledge into the primordial concepts already ingrained. For learning to take place efficiently, students must be shown reasons for evolving, for the need to seek new knowledge. This means establishing a dialectic between experimental variables and replacing so-called static and closed knowledge with open and dynamic knowledge. Common knowledge would be an obstacle to scientific knowledge, since this is abstract thinking.

Because of the difficulty students have in migrating from the macroscopic to the imagined, they can establish incorrect analog relationships when the limits of each analogy are not well defined. Differently, the teacher understands what is model and is able to migrate easily from the macro to the micro, thus establishing limits for analogies and therefore mistakenly believes that the student also has this understanding. Souza *et al* (2006), demonstrated that analogy is not always used appropriately, much less understood by students, as most of them do not recognize analogies as such; do not recognize the main analogical relationships existing in each of them; do not identify limitations of analogies; do not perceive their role in teaching; do not understand that they refer to different atomic models and do not distinguish and characterize these models correctly.

It does not seem that the atom of physics, modern chemistry and its "colligative" characteristics can be understood without evoking the history of its images, without taking up realistic and rational forms. Explaining the different models is important in the construction of the student's knowledge about the structure of the atom and the different types of atomic connections, but it is necessary to be very careful so that the necessary ruptures occur, that is, so that the explanation occurs by building a line of reasoning that leads to real learning.

What is needed, therefore, is to increase the concern with language, models, metaphors and analogies can be used, but in such a way that it will facilitate the understanding of concepts, and in the use of these "artifices", that they are not mere substitutes or simplifiers of scientific knowledge, because this practice can lead to a distorted view of science and the solidification of numerous epistemological obstacles to learning.

Melo and Neto (2013) discuss that:

> The inadequate conception of a classroom model is observed both in students (Maskill and Jesus, 1997) and in teachers active and in formation (Melo, 2002). This author soon found in his research that only 18% of the professors interviewed conceived the atom as a scientific creation, and that this percentage included master professors from a public university in São Paulo (MELO and NETO, 2013, p. 113).

When the teacher uses a resource such as models and these approaches are made inappropriately, it may not arouse interest in understanding the phenomenon and create "locks" for scientific knowledge.

These are obstacles (locks) that teachers should be aware of, so that they are not present in their way of teaching, in the school environment and in the teaching resources used, such as models. The teacher must also be aware of what each one is dealing with, as only then can he or she identify them, or help their students to overcome them, if the obstacles are present in themselves. Because of the diversity in the classroom, it leads to different rates of assimilation of the information provided by the use of these models.

When the teacher seeks to make use of a model, he should take care that it is only an auxiliary tool for the knowledge taught. And don't let that model be just a visually interesting resource, what Bacherlad calls the First Experience.

For example, in observation: the moon is not a body of light of its own, but a simple reflection of the solar fire in the aerial vault... Therefore, the stars are nothing more than the squeaky break of our visual rays, over different aerial bubbles. Reflecting better, it can be observed that a lot of importance is given in the room to images and not to ideas. Curious and fun experiences are provided, which do little or nothing to benefit the scientific culture.

Another obstacle that can be caused by the teacher's way of teaching from the use of models, metaphors and analogies, is the verbal obstacle. In this obstacle there is a tendency to associate a concrete word with an abstract one, or with an abstract model of reality. Bachelard (1996) does not oppose the use of these teaching resources in teaching, but they should be used after the theory and not before, as they should be an aid and not the main focus.

According to Barchelard (1996), the use of human attributes in science teaching can be considered a barrier to learning. Too much "embellishment", i.e. the attribution of human qualities, can lead to detachment from reality, since the model does not need to be a faithful copy of reality, but should have aspects in common between reality and the model (Mortimer, 2000), thus attributing such characteristics can lead the teacher to commit an epistemological obstacle, which Bachelard calls Animism.

Such difficulties and obstacles to teaching become more evident when trying to teach Chemical Links. Since it is observed that high school students have difficulties from the macroscopic (visible to the naked eye) to the microscopic (invisible to the naked eye) through the use of models.

1.4   Chemical Bond Teaching

We understand, according to Fernandes and Marcondes (2006), that the concept of Chemical Bond is fundamental in chemistry. The nature of chemical bonding is visualized from the electronic structure of atoms and its understanding is important for the understanding of different aspects related to the internal structure of matter and the macroscopic and microscopic properties of substances. However, the difficulties of teaching the content of Chemical Bonds are eminent. Teachers encounter not only a lack of resources or adequate structure, but the students' own disinterest.

Also, according to the authors for the student, the understanding of chemical interactions and the relationship with the properties of substances pervades the sensory level, since it requires the student to be able to move from observation to the formulation of models. Therefore, the complexity of this knowledge lies in the need for abstract elaborations, which often generate distorted conceptions about chemical bonds.

The simplified and inadequate approach of textbooks during the presentation of the content in question may lead the student to have a mistaken understanding of the concept of chemical bond (PERIZ, 2011). Others are related to the use of symbolic language. According to Maia *et al* (2007), the use of representations in the teaching of chemistry aims to facilitate the visualization of these students, regarding the formation of mental images and abstract entities, and from there also to facilitate the understanding of the nature of matter, its properties and behaviour.

The difficulty of working with this content in the classroom may be partly associated with obstacles in programming diverse didactic strategies, in addition to the lack of materials, which combine theory and experience without trivializing chemical concepts, attributing meanings closer to those accepted scientifically. Also, according to Garcia and Garritz (2006), students only recognize two types of connections as true: the ionic connections and the covalent connections. Given that the framework of work on electronic octet only provides a coherent model for ionic and covalent bonds. Students classify metallic, polar and hydrogen bonds as something other than true chemical bonds.

Another aspect that influences the student's learning is the methodology adopted by the teacher for his classes. In the teaching-learning process, the way the content is passed on can motivate and facilitate the student's understanding, at the same time as it can demotivate and cause an obstacle, making it impossible to "transpose" empirical knowledge into scientific knowledge. For Polidoro and Stigar (2006), the distance between scientific knowledge and the knowledge taught represents a transformation of knowledge that occurs in social practices, according to the diversity of the discursive genres and the interlocutors involved there. According to Lopes (2007), as one always knows oneself against previous knowledge, rectifying mistakes of common experience and building scientific experience in constant dialogue with reason, one must overcome epistemological obstacles.

It is undeniable that the rooms have become more and more diverse (heterogeneous). Thus, the current duty of teaching institutions, in the figure of the teacher, in the midst of differences, not only cultural, but also at the level of different rhythms and styles of learning, interests and abilities, in the plurality of students, is to find teaching strategies that include everyone.

1.4.1 Teaching of Chemical Connections by Models: possibilities and limitations

When you study the Chemical Links, among other contents, you see the need to use models. Mortimer (1994, p.74) "treats models not as a copy of the real but as a representation".

The use of models was initially very controversial, but throughout history it is undeniable that this procedure has contributed to the advancement of Chemistry as a Science, and as an important teaching tool to clarify ideas related to Chemical Links. It should be clear, however, that they are mere models, i.e. representations of a supposed reality that can be modified at any time (MORAIS, 2007).

An important type of scientific model that can be used to teach atomic bonds is the modeling of concrete models. Being a "concrete" resource, they allow students to create a "bridge" between the theory about the subject addressed and its visualization, as well as its "manipulation". Since to be viable the model must describe the characteristics described according to Mortimer (2000):

> As an image we build of reality and help us to understand it. In this sense, there must be aspects in common between reality and the model; a transformation that occurs in reality can be represented through the model. This does not mean that the model has to be an exact copy of reality, but that it has to represent it (MORTIMER, 2000, p.189).

Chassot (1996) considers that the choice of model, for example, atomic model should be made depending on how the shaped atoms will be used later. It is therefore necessary to be very clear how chemical bonds and electrostatic interactions will be addressed in order to evaluate the most appropriate model to be used (adopted).

For Chassot (2001), this discussion should lead the student to understand how scientific thought evolves in the face of the same reality:

> They change models, but not reality. We actually have a new atom idea, that is, a new atom, to explain a reality that has not changed. The change that occurs is in our knowledge of reality (2001, p. 259).

Thus, the use of model(s) with the purpose of teaching science, especially chemistry, while being seen as a resource that enhances teaching, is also seen as a limitation of scientific knowledge. Since analogies are used to approximate abstract models to the real world of the student, which used incorrectly can provide wrong knowledge, or that does not match scientific reality.

2. TRAINING OF CHEMISTRY TEACHERS

Teacher Training in the Latest Legislation:

There doesn't seem to be so much disagreement on the importance that the chemistry teacher's training plays for the student's learning. The teacher needs competence to encourage learning and student development. The teacher's activity is centred on monitoring and managing learning. In this sense, initial and continuing teacher training courses should have as their main proposal the training of a critical and reflective professional, who makes his classroom a place of research, where he can constantly raise questions that make it possible to advance in the use of techniques and knowledge itself.

Taking only the last twenty years as a reference, teacher training has been the subject of multiple curricular orientations and reforms, from the Guidelines and Basic Laws (1996), the National Curricular Parameters (1998), to the National Curricular Guidelines (DCN) for teacher training (2001). However, the effective outcome of curricular proposals has depended largely on how far their objective meanings established by the legislator have been respected by institutional actors and how the changes have been applied in teacher education institutions. It is easy to find, in the analysis of concrete curricular reforms, the mismatch between the original proposals, which are generally interesting and well organised, and their effective realisation, as well as cases in which there is a clear lack of respect for the legal text. There are also cases of bureaucratic application, with no investment to solve the problems encountered.

Resolution no. 2 of the National Education Council of 1 July 2015 establishes that initial teacher training courses at a higher level for basic education must have a minimum of 3200 hours of effective academic work, in courses lasting at least eight (8) semesters or four (4) years, and be distributed in this way:

I - 400 (four hundred) hours of practice as a curricular component, distributed throughout the formative process;

II - 400 (four hundred) hours dedicated to the supervised internship, in the area of training and performance in basic education, also contemplating other specific areas, if applicable, according to the institution's course project;

III - at least 2,200 (two thousand and two hundred) hours dedicated to general training activities, specific and interdisciplinary areas, foundations and methodologies;

IV - 200 (two hundred) hours of theoretical and practical activities of deepening in specific areas of interest to the students and taking into account the course project.

Such legal documents are not sufficient to guarantee the desired training, but offer support to teachers who are in favour of change, encourage those who are undecided to collaborate and reduce the resistance of those who are against innovation. Above all, they stimulate local attempts and experiences, developed by initiatives of few teachers, which try to make graduates better prepared to face the difficulties of the classroom. Moreover, the resonance between legal texts and local attempts, even when they do not result in stable changes, has had the merit of drawing attention to them, offering information on the complexity of the process of curricular change, especially from the perspective of training teachers capable of promoting the exercise of citizenship in public education.

We maintain that the effects of curricular changes do not depend solely on legal texts and are not predictable a priori, either because of the different adhesion possibilities of teachers and managers, or because of the reactions of the students. For this reason, we consider case studies, with their peculiarities, important, seeking to find common elements that involve all institutional actors or specific mechanisms that are not very visible in most cases.

From the perspective of the legislation of the National Education Council (CNE), it is possible to observe that some undergraduate courses in chemistry in Brazil had not yet concluded the implementation process of their new Pedagogical Political Projects, structured based on the guidelines present in the National Curricular Guidelines for the Training of Basic Education Teachers, indicated by the CNE, in February 2002, through Resolutions CNE/CP01 and CP02.

In July 2015, thirteen years later, the CNE established new curricular guidelines. This is RESOLUTION No. 2, of 1 JULY 2015. This resolution presents some significant changes compared to the previous one.

For example, this last resolution of the CNE, proposes that state and municipal centres of formation, as well as educational institutions of basic education that develop activities of continuing education of teachers, should consider, in their dynamics and structure, the articulation between teaching, research and extension, in order to guarantee an effective standard of academic quality in the formation offered, in accordance with the institutional plan, the political-pedagogical project and the pedagogical project of continuing education. The previous resolution deals with the issue in a very differentiated way, not mentioning extension, that is, forgetting the importance that extension can provide for learning and for the formation of the research-training-extension.

The 2002 and 2015 Resolutions state that the teacher training courses in operation should adapt their Resolution within 2 (two) years of its publication. As stated above, thirteen years later there were still courses that had not adapted the first Resolution.

> Initial and continuing training are intended, respectively, for the preparation and development of professionals for teaching functions in basic education in its stages - early childhood education, primary education, secondary education - and modalities - youth and adult education, special education, professional and technical secondary education, indigenous school education, rural education, quilombola school education and distance education (BRAZIL, 2015).

We note that, in this respect, the scope is intended for the preparation and development of professionals for teaching functions in basic education and for all forms of education. The training of teachers for higher education is excluded.

In this sense, the National Association for the Training of Education Professionals - ANFOPE, has stood out in the national scenario for fighting for the definition of a national policy for the training of education professionals, which aims at their "professionalisation and valorisation", contemplating, "working conditions", "decent salary and career and continuing training as a right of teachers and an obligation of the state".

Unlike this proposal, we have witnessed in recent years in Brazil, the opening of light graduation courses, some of them with classes only on weekends, or at night with three years duration. In general, these are Degree courses in Mathematics based basically on experiential and practical knowledge, and are required to act as trainers, teachers with extensive experience in basic education, but with little theoretical and scientific training and without the requirement of a teaching supported by research.

The National Education Guidelines and Foundations Act established that from 2007 onwards, only basic education teachers, qualified at a higher level or trained through in-service training, would be admitted. As a result of this determination, the streamlined courses referred to above have emerged, implementing the accreditation of Institutes of Education or Superior Normal Courses without any commitment to research. In order to offer degree courses for in-service teachers, distance learning programmes have also been created.

The most serious thing is that these Institutes today are offering courses in an indiscriminate way, and not only for teachers in service. The problem with these emergency projects, which as the name implies, should only be in force for a certain period of time, is that they become part of the regular practices of teacher training and even become a permanent policy.

2.1 What Chemistry Teacher Training do we want?

Nowadays, the teacher faces several challenges in his daily life, ranging from knowing how to deal with the school culture, to the explosion of information mainly coming from the language provided by multimedia. How to deal with all this by associating the socio-cultural differences so present in the classrooms? Reflection made by Chacon *et al* (2015), knowing that this new generation of learners has a greater interest in the virtual world, prefers music over reading and has access to rapid information, but often does not internalize it in a substantial way, not attributing meaning to it and thus does not transform it into knowledge.

According to Costa (2006), in the contemporary world there must be collaboration between the areas of Communication and Education, not only to facilitate the dialogue between student and teacher, but also to expand the possibilities of using resources to support teaching. In this way, we must recognize the importance of the media in the formation of the citizen. However, is the teacher prepared to work using diverse resources in his classes to motivate and instigate his students? To what extent does the use of multimedia resources favour meaningful learning?

It is not advisable to separate reflection on teaching and learning chemistry from reflection on teacher training. The training of chemistry teachers is still supported by disciplinary paradigms and the curricular structure (SANTOS, 2005).

A training model that disconnects teaching and learning chemistry from teacher training receives criticism for the way knowledge is treated, taught through an approach that favours memory, a linear and decontextualized knowledge. In Zunino's opinion (2006), this segmentation makes it impossible for the student to understand the environment in which he is inserted, to question his daily life, his reality.

Thus, the teacher is given a model in which the rationality of the technique predominates, where the final stage would be the application, by the trainee, of the theory produced in the University in his daily practice in the school, such application being carried out according to the interpretation obtained by the future teacher.

Research on teacher training has shown the need to overcome the technical rationality model. This model proposes the separation between theory and practice and promotes the fragmentation of knowledge, as well as considers science from a positivist perspective.

According to Pereira and Allain (2006), there is a need to overcome the dichotomous perspective of training, in which pedagogical disciplines are considered an appendix of scientific training. This appendix is probably linked to the idea that researchers are the intellectuals and teachers are the practical ones, the latter being responsible for the application of the knowledge produced by the researchers.

There are several problems that occur in the training of teachers in Brazil, and not only in the training of the chemistry teacher. The question of (articulation) between specific and pedagogical knowledge also involves another aspect: the pedagogical teachers who, in some institutions, are teachers who have no training in chemistry, which also makes it difficult to bring chemical and pedagogical knowledge together (SILVA and OLIVEIRA, 2009).

As we know, the objective of the Degree in Chemistry is to train the teacher to act in basic education. This training should contemplate aspects inherent to the formation of a good teacher, such as knowledge of the content to be taught, pedagogical knowledge about classroom practice, knowledge about the construction of scientific knowledge, specificities about teaching and learning chemistry. In view of this, it is necessary to take into account that initial training courses and teachers working in training "promote new practices and new training tools, such as case studies and practices, long-term internships, professional memory, reflexive analysis, problems, etc.". (ALMEIDA and BIAJONE, 2007, p.293).

Changes in the training of the chemistry teacher are desired. Some authors glimpse a new dawn. Actions to change this scenario have already begun to be practiced, with innovative proposals for the Degree courses in Chemistry (BAPTISTA *et al*, 2009).

Concern with the current reality is a constant, there is a desire for change in the formation of the teacher. The results will only tell the future its size and scope, if we are forming a teacher capable of offering possibilities different from the current ones to the student, above all, possibilities for effective learning.

We leave the University with the feeling that we are not prepared or not prepared to be teachers, to deal with a classroom composed of students, where each student learns differently. We have the slight impression that we are being "forged" into researchers, when in fact we should be teacher-researchers, i.e. we left college with the impression that Academic Education forms researchers, not teachers.

This condition to train teachers, according to Luiz Carlos de Menezes, professor at the Physics Institute of USP and member of the Scientific Technical Council of Capes, in his speech at the "Seminar: Teaching Science in Brazil - In search of new strategies", held by the Fernand Braudel Institute of World Economics in the year 2015, states that there is no teacher training in Universities and Colleges. The true formation of the professor happens when he goes to the classroom to exercise the teaching activity. The professor points out this situation as a great problem in the formation of qualified professionals in this area in the country. Since quality teaching depends on good professionals, it is necessary to bring the future teacher closer to the school and to school practice, living these circumstances.

> It is not possible to continue training a teacher for a reality different from that which he will have to face; therefore, the question of practice, in the context of the school reality of the exercise of the profession, becomes an important formative principle (RAMALHO; NUNEZ; GAUTHIER, 2004, p. 176).

We believe that in degree courses where the approach of the future teacher to the school reality, will be given merely by teaching internships, is insufficient or incomplete for the training of the teacher. Educational programmes, such as PIBID, which is an introductory programme to teaching, are of paramount importance in the training of teachers, not only to enrich their curriculum, but also to the extent that it provides graduates with an experience of living in the public schools of the educational network, it provides them with the opportunity to develop projects that will improve the quality of teaching in their field, as well as, from the work that is carried out in schools, to get to know the teaching staff and the school reality.

The teacher is one of the main actors in the educational process, whose main function is not only to teach, but also to produce knowledge in a constantly changing society. As we move from primary to higher education, teachers value scientific skills to the detriment of pedagogical skills.

To this end, the teacher-researcher profile is increasingly necessary in a changing society, and this function is complementary and facilitates conceptual changes about the teaching activity, which cannot limit itself to transmitting knowledge, but in a broader vision, seeking ways so that more teaching methodologies can emerge and enrich their teaching practice.

How do we train the science teacher, for the teaching of chemistry if all he experienced during his academic training was as a spectator, being a mere listener to the exhibition of theoretical aspects? There must be a more "intimate" relationship between Academic Education and Basic Education. In order to strengthen the links between the educator's formation and the school reality of schools, especially public education institutions, where the educational reality is often "chaotic" (its disturbance not in the grammatical meaning of the word), either due to the lack of an adequate physical space, or even the lack of security for the teaching exercise.

We believe in the teaching process and in the occurrence of learning in public schools, but we need, just as it is possible, to offer a better quality education, and without reservations the more it is invested, either financially, or in the development of new methodologies for the formation of professionals with better qualifications, the more positive results there will be in education. The teacher must be able to awaken the student's interest in the subject addressed. This is not an easy task in the midst of a generation, it leads to dispersion, the lack of attention in the classroom for the many attractions that technological advances offer.

In this sense, class is not always synonymous with joy, because there are no magic formulas to teach how to teach. The classroom can become a place of teaching frustration when students do not value you, do not respect you, do not pay attention. What exists, however, are methods that serve a certain situation, student, teacher, but that are not a cake recipe, that is to say, the teacher himself from his training and with the practice in class will allow them to develop their own teaching methods. We need to understand that the methods, strategies and resources used, serve as a support for the performance of the teaching activity, but the teacher himself is the best "methodology".

Therefore, formative practices, as well as teaching practice, will be better understood and/or developed from the awareness that the human being is a complex being. A multifaceted, social and biological individual.

3. THE TEACHING SEQUENCES

One way of working with chemistry teaching is through the adoption of a proposed didactic sequence. This is justified by the way this type of work is organized, aiming at contextualizing the content presented. In our study understanding the meaning and significance of a didactic sequence is essential. In this sense, we begin to discuss its meaning, presenting the interpretation and definition of some authors.

The Teaching Sequence is a set of linked activities, planned to teach a content, step by step, organised according to the objectives that the teacher wants to achieve for the learning of his students and involving evaluation activities that can take days, weeks or during the year. It is a way of fitting the contents to one theme and in turn to another making the knowledge logical to the pedagogical work developed.

According to the Didactic Sequence scheme presented by Dolz, Noverraz, Schnewly (2004), teaching takes place initially by presenting a situation, regarding the content addressed. This is followed by a survey of the students' previous conceptions on the subject, called here the initial production. The interventions carried out are called modules, and finally the final production is carried out, which seeks to demonstrate the evolution in student learning in relation to the proposed content.

According to Araújo (2013), the didactic sequence model is related to research on the acquisition of written language through work developed with textual genres by a research group in Geneva (Switzerland). Representatives of this group, Dolz, Noverraz and Schneuwly (2004), define Didactic Sequence as "a set of school activities organized, in a systematic way, around an oral or written textual genre" (DOLZ, NVERRAZ and SCHNEUWLY, 2004, P. 97).

As proposed by Dolz, Noverraz, Schnewly (2004), the teaching takes place initially by presenting a situation, regarding the content addressed. This is followed by a survey of the students' previous conceptions on the subject, called initial production. The interventions carried out are called modules, and finally the final production is carried out, which seeks to demonstrate the evolution in student learning in relation to the proposed content. Araújo (2013) summarises what is a didactic sequence thus: "it is a way for the teacher to organise teaching activities according to thematic and procedural nuclei" (ARAÚJO, 2013, p. 3).

For Kobashigawa et al. (2008), this is a set of activities, strategies and interventions planned step by step by the teacher so that the understanding of the proposed content is reached by students. Something similar to a lesson plan, but with a wider scope, as it addresses various teaching and learning strategies and is a multi-day sequence.

In the opinion of Zabala (1998), the didactic sequence is a set of ordered, structured and articulated activities for the achievement of certain educational objectives, which have a beginning and an end known by both the teacher and the students. For this author, didactic sequences can be understood as a way of situating activities, which cannot be seen only as a type of task, but as a criterion that allows preliminary identifications and characterizations in the way of teaching.

The socialization of experiences related to the teaching and learning of chemistry in the classroom can create possibilities through collaborative actions between students and teachers, which will favour concrete and real work in the construction of pedagogical practices. "We must insist that everything we do in class, no matter how small, focuses to a greater or lesser extent on the training of our students" (ZABALA, 1998, p. 29).

According to Vargas and Magalhães (2011), a Didactic Sequence is a set of systematized pedagogical activities, linked to each other, planned stage by stage, with the aim of the mastery of a certain oral or written genre by the student and the development of his/her language skills.

For Ayres and Arroyo (2015), the Teaching Sequences are able to offer opportunities for building relationships between teachers, students and content. According to the role assigned to each one within this process, we will have an effect, a consequence for the planned activities and, consequently, for the didactic sequences. According to these authors, the diagnosis is essential in the application of a didactic sequence, because it allows the teacher to recognise what the student understands about the content, favouring the teacher to adapt the development of the activities.

This is a method widely used in school education to work on certain contents with pupils, as well as in the teaching of chemistry. It is a methodology that actually helps the teacher a lot in the planning of his classes and allows the students to have a wider view of the contents. Since the Teaching Sequence is elaborated and developed in terms of a central content, at all times the subjects dialogue with each other.

When using the Didactic Sequence, for example, in the teaching of Chemical Links, the teacher will necessarily perceive the need to approach themes such as, Periodic Table, Octet Rule, Valencia Layer, Electronic Distribution, among other contents of Chemistry, once these themes are linked.

When we talk about Didactic Sequence, two similar but distinct concepts occur in its planning and execution. In the first concept every teacher, when going to class, has planned what he is going to do with his students, in this perspective every class has a sequence to follow, gaining in this sense a connotation of sequenced activities, that is, there is the organization of the activities that will be carried out in a shorter space of time. Of course, the time allocated or the quantity of activities carried out are not what characterises the viability or significance of the didactic sequence, but rather the planning, execution, purpose, involvement of both students and teacher.

The other concept occurs when we talk about Didactic Sequence as a pedagogical instrument in the educational formation of students. We are talking about an instrument for organizing the teacher's time and space that is smaller than the execution of a pedagogical project, but which, unlike the simplicity of "sequenced activities", requires a greater amount of time in the organization, planning, development, and execution of the didactic sequence. That is why the sequence should be thinking together "where I am and where I want to go". In this way "what am I going to do", "where am I going to accomplish", "how many lessons will I spend" to reach my goal.

According to Landeira (2016), the Teaching Sequences have been divided into different parts: problematisation, development, synthesis, reflection and transmutation.

In research, problematization is the essential point, the conductive part by which the researcher will develop his work. Thus, to problematize means to question a given object from the interests of the researcher. In the classroom, the researcher becomes the teacher and the research becomes the didactic sequence. Thus, the problematization in the didactic sequence, we try to leave in evidence to the student the problem of the study, that is, we expose the problem that apparently there is no solution. Once we have the immediate answer to this one, it would not be necessary to use the Didactic Sequence.

In development, as the definition itself is the moment when the teacher establishes his or her strategy(s) to apply and develop the activities that will be worked on through the didactic sequence. In the synthesis, the ideal is the appropriation of the learning by the student, that is, the systematization of the concepts that will be effectively learned by him/her. At this stage it is very important that the student has the ability to catch up with the processes in the teaching and learning path.

The next stage, reflection, is important that the student awakens the capacity for criticality, for enquiring about the path taken, the path followed. What skills and competences have been developed? In fact, was participation and involvement effective? From this condition the student starts to tend to the importance of each stage followed in the didactic sequence, and

through the instant that the concepts gain meaning for the students as they are submitted to more active and stimulating learning methods, thus teaching approaches that transform students into active participants reduce failure rates. It is important to enable each stage of the activity to be developed to favour the use of models as a means of mediating the theory described with the "visualisation" of it, providing conditions of understanding on the subject addressed.

The figure of the teacher is very important in this process, since the Didactic Sequence is a means of organising the pedagogical work, allowing us to anticipate what will be focused on in a space of time that is variable according to what the students need to learn from the mediation and the constant monitoring that the teacher does to accompany the students throughout their journey, that is, from the development, application to the evaluation of the students. The teacher not only transmits knowledge but also acts as a mediator of knowledge. He or she must act as a bridge between the student and the knowledge so that the student awakens his or her critical sense and begins to question the world around him or her by himself or herself, not passively receiving the information. It is therefore important for both the teacher and the student to know all the comprehensive stages of the didactic sequence.

Some assumptions are considered obstacles to being overcome so that the teaching exercise and the student's need can walk together the educational strategies that oppose a merely passive and accumulative practice, and why not say quantitative. In view of the usefulness in employing the didactic sequence(s) as an intensifying "agent" in teaching and facilitator in learning, the educational system of public schools does not collaborate in the use of this method. The number of students per class, the number of students in a class, and the curricula are many factors that limit the use of this didactic in the classroom. Among the main difficulties are the short time available to carry out the activities in the midst of a didactic sequence, the indiscipline of the students. In this sense, it is necessary to rethink this articulation with the current workload for the discipline of chemistry, the possible lack in the training of teachers to act with this modality, the excess of contents to be taught, and the need for the teacher to reconcile theoretical lessons and experimental activities, which is a very important aspect in the process of teaching and learning the contents, without this being reflected in overload of work.

From the analysis of the research carried out by Reis (2013), the lack of knowledge of the teachers, targets of the research, to work with the Didactic Sequence(s) in their classes is evident. This limitation may occur due to the "bad" academic training of these professionals, since many teachers leave universities without a training on the importance of using this teaching methodology. It is also possible to highlight a misconception or misconception for use in basic education.

Kings (2013) discusses that:

> An important point raised in the questionnaire was in relation to teachers' conception of a didactic sequence for teaching Chemical Links through the use of models. In the light of teachers' responses we found that the vast majority do not understand the concept of didactic sequencing, so this can be seen as an obstacle as they will not be able to develop planned activities that are able to generate solid learning through potentially significant didactic sequences (REIS, 2013, p. 52).

Also, according to the author, it was sought to identify whether teachers believe that the use of the model in isolation, i.e. in that context, without the intervention of a didactic sequence allows students to have significant learning.

> It became clear that a large proportion of teachers in their responses highlighted the need to work on the use of models through a well-planned teaching sequence. Thus, 22% demonstrate that they do not need to use teaching sequences, which leads us to understand that these people possibly adopt in the classroom a teaching from a traditionalist perspective. This is evidenced by the DCNEM when they state that the Teaching of Chemistry opposes the old emphasis on memorizing information, names, formulas and knowledge as fragments disconnected from the reality of the students. On the contrary, it wants the student to recognise and understand, in an integrated way and significativa, the chemical transformations that occur in natural and technological processes in different contexts (REIS, 2013, p. 50).

According to Reis (2013) the lack of teacher training in understanding and planning didactic sequences of teaching is an obstacle, however, it is perceived that the use of the Didactic Sequence, as observed from the "traditionalist and limited" view of the teachers targeted by the research, in addition to the obstacles mentioned, the use of this resource can be a very useful tool that will contribute to the teaching-learning process of the contents of chemistry.

4. METHODOLOGICAL ASPECTS

4.1 A Qualitative Research

The literature is very diverse as to how to classify research. In order to understand the phenomenon of study, we perform a qualitative approach, as a strategy of apprehension of the object of research, it also presents a research character.

Regarding qualitative research, André (1995) states that for some it is phenomenological research. For others, qualitative research is synonymous with ethnography. For still others, it is an "umbrella term that may well include clinical studies". And at another extreme, there is a popularized idea of qualitative research, identifying it as one that does not involve numbers, i.e., in which qualitative is synonymous with non-quantitative.

For Richardson (1999), qualitative research can be characterised as an attempt to gain a detailed understanding of the meanings and situational characteristics presented by interlocutors, rather than producing quantitative measures of characteristics or behaviours.

The definition of qualitative research, for Richardson (1999), poses several problems and limitations. First, few attempts are made to place the conceptions and behaviours of research interlocutors in a historical or structural context. It is considered sufficient to describe different forms of consciousness without trying to explain how and why they developed.

According to Richardson (1999), this leads to a second problem, the tendency to adopt a non-critical attitude of the conceptions and consciousness of the interlocutors of research, without considering their epistemological development.

The characteristics of the qualitative method are present in several authors. The six main ones, which are specially defined by Lincoln and Guba (1985), Miles and Huberman (1994), Lüdke and André (1986) and André (1995), are highlighted below.

1)   The researcher is considered a research instrument, which can use his experiences, his tacit knowledge and his existential assumptions to collect the data, understand them and interpret them.

2)   The qualitative approach presents descriptive data that are addressed interpretively. They are collected in the form of words that seek to translate as much as possible as things have happened. It usually contains literal quotations, figures and other resources that help to reconstitute the scene investigated in order to provide a "holistic" view of the research context. The data tend to portray experiences as they are "experienced" by research participants, seeking to translate the way they structure, perceive and give meaning to them.

3) The natural environment is the direct source of the data. It refers to the situations where research occurs, whether current or arranged. Qualitative research requires prolonged contact with the field in which the research is carried out. It is through this attempt at insertion into the environment of the research participants that one can describe and select the aspects judged central to individuals.

4) The understanding of the process has a relevant place for qualitative researchers, who want to know how phenomena occur from their internal characteristics.

5) The search for meaning that people give to things is the central point of qualitative research. As knowledge of reality is perspective, that is, it is given by diverse perspectives, it is important to bring the subjective point of view to the understanding of reality. The meaning concerns the way people designate, translate, interpret or intend recaptured experiences.

6) The method of analysis is inductive, so that one does not work with any theory or hypothesis a priori, but seeks understanding from the data. This does not mean that the researcher enters the field unloaded from his assumptions, but rather that they interfere in the conduct of the research. Nor does it mean that there is no theoretical framework to support the collection and analysis of data. What has not been established beforehand is a theory - a set of laws and definitions - which generates hypotheses to be empirically verified. The inductive posture opens up the possibility of creating new theoretical concepts instead of "confirming" a previously established theory.

According to Moreira and Calefe (2006) qualitative research explores the characteristics of individuals and scenarios that cannot be easily described numerically, so the data is often verbal and is placed by observation, description and recording.

In this sense, this research is qualitative and based on the analysis of the performance of students in the 1st year of high school, after the application of a Didactic Sequence based on the use of molecular models in the teaching of Chemical Bonds.

4.2 Characterization of Research Interlocutors

Two (2) late shift classes were selected. An average of 64 (sixty-four) students in total. The students from the school chosen are students who mostly do not live near the school, since most of them live in rural areas. The class sizes are around 38 students, however classes are usually for 30 to 34 students. Students coming from primary school are either from the school itself or from the municipal school of education, some bring difficulties from primary education such as problems of reading, writing and mainly mathematical calculations.

For the application of this Didactic Sequence, students will be from the 1st year of high school of the late shift, since the rate of passing or even acceptance of students for chemistry has been very low in recent years. There is also a personal reason for choosing the afternoon shift for this research. Although the students who study in the morning have a higher pass rate, as this is reported by the school's board of directors who make an annual record of students who pass courses in higher education institutions. Therefore, we are led to think that they would possibly have more involvement in the application of the stages of this research, however, it has been 7 (seven) years of study from grade 5 to the 3rd year of high school, during this period a bond of friendship and respect has been created with teachers and other employees.

In addition, my "entrance" to higher education was gained from the teaching-learning process acquired in the afternoon shift, despite the "prejudice" and sometimes "debauchery" on the part of some... who did not believe in the ability of these students to enter a university, as most of the students "are from the place".

4.3 Characterization of the School

The State College of Elementary and Secondary Education of Alagoa Municipality Nova-PB, located at Travessa Maria Lima Maracajá, 85, has approximately 1500 students from 6th grade to 3rd grade and operates in the morning, afternoon and evening shifts. The students come from the municipality's schools, located in rural and urban areas.

In 2016 the school approved 26 students who were invited to enroll in Higher Education Institutions - SISU 2016.1 and 2016.2, such as: UEPB, UFCG and FACISA. Among the approved courses (Biological Sciences, Physics, Mathematics, Nursing, Dentistry, Law, Sociology, Meteorology, Social Sciences, History, Biological Sciences, Letters/English/Spanish), no course is aimed at teaching Chemistry.

4.4 Data Collection

This is a qualitative survey, with the application of a pre-questionnaire composed of closed questions, which contained information concerning the objective of the survey in order to have a prior analysis of the students' knowledge(s). We request a textual production of the students on the subject addressed. Thus, as a final written activity from the molecular models elaborated by the students representing the types of Chemical Bonds.

We also used a field diary for data collection, where throughout the research the statements of some students and teachers that will be used in this work for analysis were recorded. And we made the recording of images during the application stages of the Didactic Sequence.

4.5 Data Analysis

In the subjective questions (textual production and final activity) discourse analysis was used and the objective questions (questionnaire) were placed by means of graphs and charts, and later discussed.

We also use discourse analysis for the statements of students and teachers recorded during the application of the Didactic Sequence. Also, in the final activity we build a framework where we perform the synthesis of the students' responses.

In addition, the images recorded of the students during the teaching sequence activities were discussed.

5. RESULTS AND DISCUSSION

Society is moving towards a "social state" where the valuing of diversities and scientific advances require the school, the innovation of its pedagogical practices. In the field of education as important as studying the "protagonists" involved (teacher and student) it is necessary to investigate the different teaching themes, whether the strategies and/or teaching resources used in the schooling process.

Starting from the question, "who is being taught", we are led to reflect on cultural plurality as well as cognitive diversity, where each individual (the student) has his or her particular way/time of assimilating the contents. In this way, we start thinking about other questions, such as "what to teach", "how to teach" and "teach for what".

The following data refers to the 6 (six) classes held with the students of the 1st Year of the late shift of the State College of Elementary and Secondary Education in the municipality of Alagoa Nova-PB. It consists of the stages of the Didactic Sequence applied with the 2 (two) classes, about 68 students, and will be subject of analysis and discussion for this academic work.

In the chart below it is clear, and we can relate the affirmatives with the marked alternatives, where we can check the 34 (thirty four) possible alternatives to be marked for each alternative: RIGHT; WRONG or NON-REMBER.

Figure 2: List of affirmatives with marked alternatives.

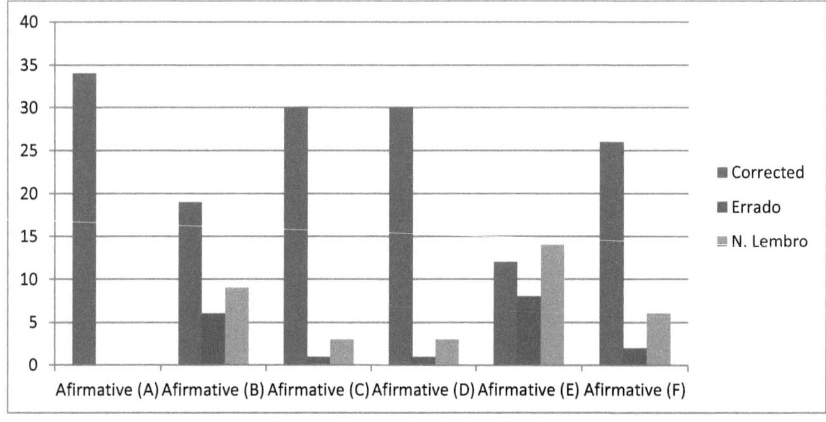

Source: Researcher data collection.

Of the 6 (six) affirmatives, 5 (five) present the alternative: CORRECT, as most pointed out by students.

The 1st (first) affirmative: *"a) Chemistry is a branch of Nature Sciences that studies matter, its properties, constitution, transformations and the energy involved in these processes"*, was pointed out as being correct by all students.

It is one of the first concepts studied by students in the 9th grade in Science Teaching. However, (mention teacher training) and mention the speech teacher A "Teachers in 9th grade when they go to teach, since science: Chemistry, Physics and Biology, are seen together, the teacher tends to pull into his or her area of training". For example, the teacher who has a degree in Biological Sciences tends to go deeper into concepts of Biology.

Already in primary school in the initial year, since the initial chapters of the textbook "Being Protagonist - Chemistry 1", used by teachers of chemistry in their respective college to teach the classes in the 1st (first) affirmative year, address the concepts mentioned in the 1st (first) affirmative. Therefore, the students' familiarisation with the concepts are more recent if we take into account the beginning of the school year starting in February and the application of the Teaching Sequence starting in April.

According to Professor B:

<blockquote>The sequence of contents of the book is not necessarily followed, but an overview of the subjects is sought, where we understand that chemical concepts (matter, energy, substances, properties and transformations) should or should be seen simultaneously (Field Diary, April 2017).</blockquote>

From the graph (figure 2) we can see that the alternatives: WRONG or NON-REMBER, appear with greater markings in the affirmatives:

*b) Two forces of different natures are observed in chemistry; the **intramolecular, which** occur within a molecule, and the **intermolecular**, which are interactions between two or more molecules, equal or different.* Where 26.5% of the alternatives were marked as: NON-LEMBER, and 17.6% of the alternatives were marked as: WRONG, by the students.

And in the affirmative:

d) According to the grouping, the substances have distinct properties, so Chemical Bonds are classified according to these properties. 23.5% of the alternatives were marked as: WRONG, and 41.2% and alternatives were marked as: NON-LEMBER, by students.

We can interpret according to Professor A.'s speech:

<blockquote>Often the problem, the difficulty of the student(ies) is not in the scientific concepts, be it chemistry or other natural sciences, but in the contextual interpretation, and the operation of sum, subtract, divide and multiply mathematics (Field Diary, May 2017).</blockquote>

As we noted only in the 5th (fifth) affirmative, which was constituted in: *"According to the grouping, the substances have distinct properties, so Chemical Bonds are classified according to those properties"*, the alternative: NON-LEMBER, was the most marked.

Interesting given that in the 4th (fourth) affirmative, which said: *"Substances are made up of groupings of atoms or ions that bind through Chemical Bonds"*, where more than 88% of the alternatives were marked as, CORRECT, but the students could not correctly mark the later affirmative. When attributed a greater degree of complexity in the interpretation of the 5th affirmative, we noticed that the students manifested uncertainty whether the affirmative is correct or wrong. In this sense, the respondent may be influenced by the alternatives presented.

From the graph below, we summarize in a general way the percentage of the three alternatives marked.

Figure 3: Percentage of the alternatives marked.

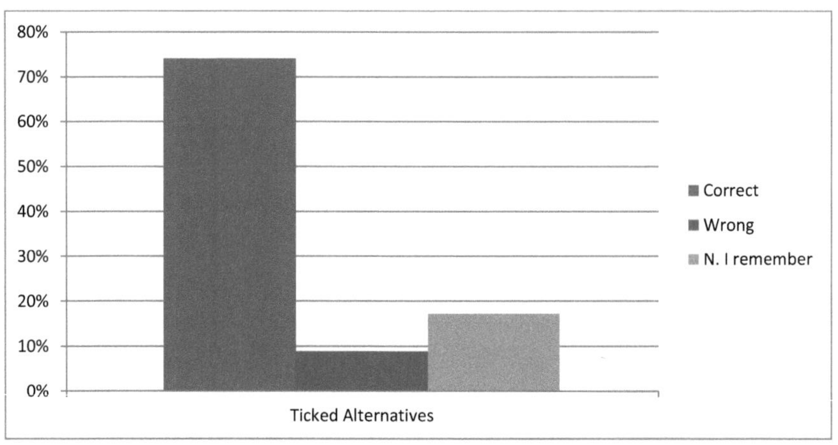

Source: Researcher data collection.

From the application of these statements we aim to know the students' previous knowledge about the content. This is an important factor discussed in Meaningful Learning.

We can see from the graph (figure 3) that most students have previous knowledge (or at least some understanding) about the content of Chemical Links. Of the possible alternatives to be marked for each statement, i.e. all statements could be marked as "CORRECT", "WRONG" or "NON-CORRECT", we can see (Table 1) that students marked 74% as "CORRECT" for the statements. Therefore, most students have previous knowledge about the content, it does not mean that they scientifically understand the Chemical Links, but that they present a knowledge, and this knowledge can, and should, be used for the application of this activity.

According to Mantovani (2015):

> [...] meaningful learning occurs when ideas expressed in a symbolic way interact in a substantive (non-literal) and non-arbitrary way with what the learner already knows. Such interaction does not occur with any previous idea, but with some relevant knowledge already existing in the cognitive structure of this apprentice... In other words, in a simple way, subsumption is the name given to all specific knowledge existing in the apprentice's knowledge structure, which allows him/her to give meaning to a new knowledge presented or discovered by him/her. Non-arbitrary implies the relationship of the new knowledge to a specifically relevant knowledge (subsuncture) and not to any other existing knowledge in the learner's cognitive structure (MANTOVANI, 2015, p. 11).

Another important fact is that in only three (3) of the questionnaires, all the statements were marked "CORRECT" by pupils. This is a comforting, but at the same time alarming sign, as it is a fact that many students entering high school have great difficulties in learning basic

44

concepts of chemistry, partly because the classes taught by the teachers are still merely expository and free, without the use of demonstrations and/or experiments related to the theoretical content taught. Or even by the lack of interest of the student who cannot relate the "mechanized" content with the environment where they live.

These possible difficulties, specifically the fact that many students are unable to relate the content studied to their daily lives, is also observed by Batista *et al*, (2013), when he seeks to raise in his research whether students, before the application of the didactic sequence, used chemical concepts to solve, interpret or understand a practical situation in their daily lives.

Batista *et al*, (2013), discusses:

> That 17% of students use the knowledge of electrochemistry content in everyday situations. Already 83% do not use this knowledge in everyday situations. The PCNEM (BRAZIL, 1999, p. 138) discusses that contextualising the content to be worked on in the classroom is an important resource for taking the student out of passive observer status and making learning meaningful by associating it with everyday life experiences or with knowledge acquired spontaneously (BATISTA *et al*, 2013, p. 1-12).

Taking as a starting point the exposition of what would be worked, the steps to be followed in the Didactic Sequence, the questionnaire applied and through the observations made on the behaviour of the students, we proceed to the second class.

SECOND (2ND) CLASS

We have continued to address the issue. To start the class, we used the questionnaire applied in the previous class, explaining the statements. The image below illustrates the moment during the class when we tried to clarify the statements in the questionnaire.

Figure 4: Explaining the statements in the Questionnaire.

We believe that students who have undergone traditional classes are more likely to fail than students who are in contact with more active and stimulating learning methods. We seek to establish a dialogue with students where they can participate, as we observe that teaching approaches that turn students into active participants reduce failure rates.

From this perspective we believe that the use of technological resources can make classes more attractive. Since we use "audiovisual" to catch the students' attention. In an increasingly computerised and globalised society, the use of computers, among other technological resources, has been playing an increasingly important role in people's daily lives and in education. Therefore, it is not possible to think of a teaching-learning process that does not integrate technological resources and educational practice.

Therefore, the teacher needs to understand the changes (technological advances) to act as a mediator between the technologies used in teaching and the students' learning. Without a doubt, the use of technological tools makes for a more "attractive" class, where the relationship between content and student, teaching and learning is enhanced. But the teacher must be aware that they will not replace his or her work, as it is he or she who will plan the classes and know the best moment and the best technological resource to complement a certain content. In this way the technological resources should be seen as a teacher's supporting tool.

In an increasingly "learning" contemporary society, the use of technologies in teaching is a very current concept among researchers, as well as their use in the pedagogical practice of some teachers. In the midst of scientific and technological advances and the proposals of new teaching themes, for example, CTSA education (that the Science, Technology, Society and Environment teaching approach is linked to the scientific and environmental education of the citizen), it is necessary to engage in a "more comprehensive education", i.e., the realization of activities that address this theme, as well as the adequate provision of teaching resources that can be used by teachers and students in the teaching-learning process.

That is why our responsibility as educators is to undo the mysticism that exists among students about chemistry, where we observe that chemistry as a subject is poorly seen and understood by students. This is a very strong concept among the students.

According to student C:

> *Professor, I don't know anything. I don't know any chemical process. I don't know how the teacher likes it. Chemistry is very difficult (Field Diary, April*

2017).

Therefore the need to use teaching methodologies that provide effective learning, once they have the ability to be facilitators and more attractive to students.

According to Silva Pereira, et. al. (2013) in a research presented at the third meeting of the UEPB for the initiation of teaching, it seeks to highlight the importance of the didactic sequence for the approximation of the students' real scientific knowledge and to promote a reflexive teaching. One aspect raised was to analyse whether the students can understand the importance of chemistry in their daily lives through the contents worked by their teacher.

Batista *et al* (2013), he says:

> That 67% of students are able to perceive the importance of chemistry in their daily lives. While 5 % do not perceive any connection with their everyday life, and 28 % sometimes do. Although most students perceive the importance of chemistry in their daily lives, many teachers still work in the classroom from a traditionalist and out-of-context perspective, working so that 33% of students do not understand the role and importance of learning chemistry in today's society. As Chassot states (2003, p. 126), "chemical knowledge, as it is usually transmitted, detached from the reality of the student, means very little to him" (BATISTA *et al*, 2013, p. 1-12).

In this sense, there is a need to overcome the current teaching practice, providing access to chemical knowledge that allows the construction of a less fragmented and more articulated world vision, contributing to the subject seeing itself as part of a world in constant transformation.

To continue the lesson, we asked students to discuss and write down in the notebook what they knew about Chemical Links. We then intervened by presenting the concept of Chemical Connections, the types and characteristics of each one. To make this explanation, we used the Datashow to present the students with some molecular models used to illustrate the concept of Chemical Bond and the different types. The image below shows some of the models used.

Figure 5. Models used to show the characteristics of Chemical Bonds.

Figure 6: Textual production developed by G/H students.

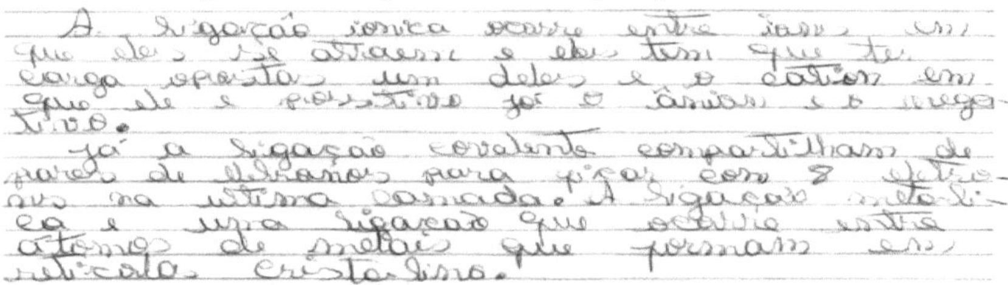

Source: Researcher data collection.

From the text produced by the students we observe the evolution in learning through the Didactic Sequence followed. We notice that the students are able to conceptualize the three types of Intramolecular Chemical Links approached in class. We also notice that other chemical concepts are presented by the students in the characterization of the links, such as "Rule of the Octet" and "Valencia Layer". From the students' response it would be possible to work on other contents, such as Electronic Distribution, Chemical Reactions.

Two other texts produced drew attention to the fact that students cited bronze and gold as examples of metal bonding. As well, the ability to make a "rescue" of content to define the concept of Chemical bonding.

> *One atom. What is an atom? An atom is a microparticle that is divided into protons, neutrons and electrons. Protons are positive particles, electrons are negative and neutrons are neutral... (Students M/N).*
> *A metallic alloy, bronze, gold they can be types of metallic bonds... (O/P Students).*

Through the enquiry made by the students (M/N), we can observe their critical sense. Their ability to "think critically" about the subject addressed for the formulation of their answer. Competence that we seek to awaken in students.

> Since the mid 1980s and during the 1990s, science teaching has challenged active methodologies and incorporated the discourse of training the critical, aware and participatory citizen. The educational proposals emphasized the need to encourage students to develop reflexive and critical thinking; to question the existing relationships between science, technology, society and the environment and to appropriate relevant scientific, social and cultural knowledge (DELIZOICOV and ANGOTTI, 1990 apud NASCIMENTO et al, 2010, p. 232).

In this sense, the questioning is necessary, that the students are involved and realize that the understanding on the subject approached occurs with the question (enquiry) accompanied by the answer (opinion) of the same, although erroneous, but the "self-criticism" is important. This behaviour should exist for the various contents of chemistry, such as Chemical Links.

Figure 7: Textual production developed by J/L students.

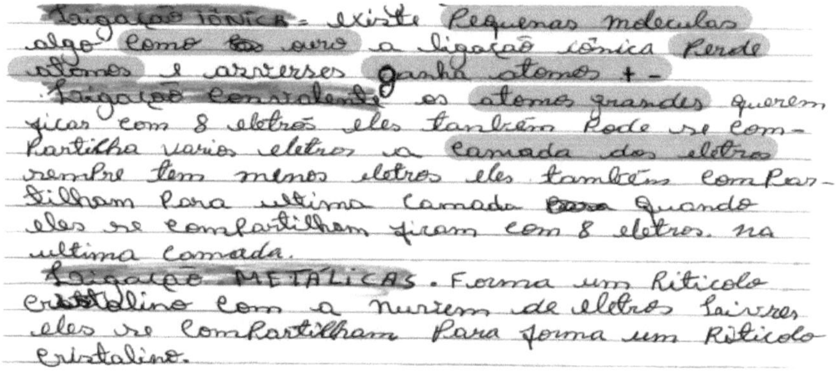
Ligação IÔNICA = existe pequenas moleculas
algo como ouro a ligação iônica perde
átomos e arverses ganha átomos + -
Ligação Covalente os átomos grandes querem
ficar com 8 eletrôs eles também pode se com-
partilha varios eletros a camada dos eletros
sempre tem menos eletros eles também compartilham para ultima camada quando
eles se compartilham ficam com 8 eletros. na
ultima camada.
Ligação METÁLICAS. Forma um Riticolo
cristalino com a nurtem de eletros livres
eles se compartilham para forma um Riticolo
cristalino.

Source: Researcher data collection.

From the text produced, we can see that students can measure some chemical concepts and even bring a definition of chemical bond, but they confuse atoms with electrons. Assigning the molecules the gain or loss of atoms to acquire stability according to the octet rule. It is also observed that they cite the golden atom to conceptualize the Ionic Bonds, since the golden atom being a transition metal participates in the formation of the metallic bonds. In this sense, it is believed that the students have been induced to this response since the textbook "Being a Protagonist - Chemistry 1", used by the chemistry teachers in the respective college to teach the classes in the first year of high school, brings in its definition, as well as, was also approached in the presentation and speech with the students the formation of a crystalline reticulum, both in the ionic and metallic bonds, as shown in Figure 5.

Another example of this possible "misunderstanding" can be seen in more texts where students associate the formation of a crystalline reticulation with ionic and metallic bonds, attributed is a characteristic example of chemical bonding.

> *Another type of ionic bond and metallic bond are those in which the nuclei unite to form a crystalline reticulum forming a cloud free of metal 'conatamos' (H/I Students).*

We observe that the students point the "union between the nuclei" to the formation of a crystalline reticulum (characteristic found in both types of connections) forming matais free

Figure 9: Students making the molecular models.

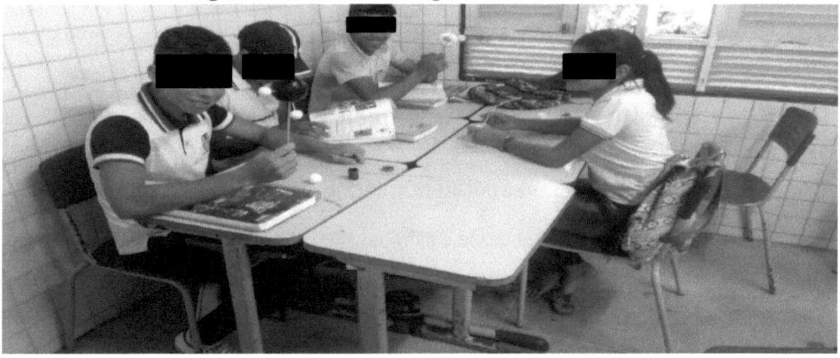

Source: Researcher data collection.

From this activity carried out with the students we can observe their involvement during the making of the models. In teaching, the use of models allows the construction of scientific concepts, when it is worked in a constructivist perspective. Therefore, its use favours the understanding of concepts that in most cases are considered difficult by students. Therefore, those who think that the role of the teacher is just to "pass on" knowledge may see active and interactive learning as a theoretical reverie or as illusions of certain pedagogical proposals.

Figure 10: Clarifying students' doubts about the representativeness of models.

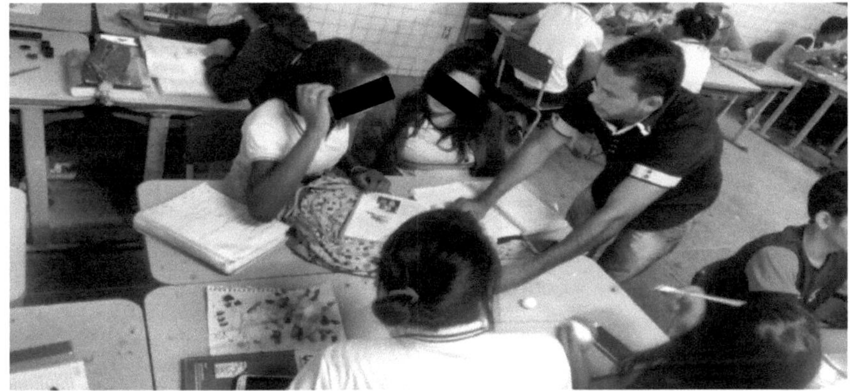

Source: Researcher data collection.

The figure of the teacher is essential in this process, because in the use of this material, one must take into account what the student already knows, so that the use of it does not have an opposite effect, thus causing an epistemological obstacle.

Thus, we must be careful not to make a superficial appropriation, of the concept of significant learning, so that the teaching strategy used, in this case, the didactic sequence, has Significant Learning not only as a theoretical objective, but gains practicality and real applicability in the classroom. For, in practice, most of these strategies continue to promote mechanical learning, purely memoristic, rather than significant.

According to Moura and Morretti (2003) an impacting aspect in learning is what the student already knows (previous knowledge). Therefore, this should be ascertained and the teaching should start from this data.

From this vision, we understand that for Meaningful Learning to occur, it is necessary that the teacher uses a good teaching methodology, which can be easily understood by the student, and makes use of a potentially meaningful material, in this particular study, the Didactic Sequence, where we make use of models not in isolation (without there being a planning for its use during the explanation of a content), but with a helper tool together with the Didactic Sequence.

Figure 11: Students making the molecular models.

Source: Researcher data collection.

Promoting dialogue (the discourse of the topic) among students by initially leading them (because the final intention is to awaken the students' criticality, their autonomous capacity to question the world around them) about the content addressed is not an easy task. Constantly the parallel conversations and the lack of attention promoted by the use of social networks (fecebook, whatsapp, twitter, instagram, etc.) hinders the teacher's activity in the classroom.

Besides the difficulties mentioned, situations such as those described by Rafael Henrique Moreira in his master's research on the teaching and learning of specific astronomy topics from the proposal of a didactic sequence with the use of diversified resources carried out in 2015, are other obstacles faced sometimes by us teachers in the classroom making it difficult for us to teach and learn from other students when they witness such manifestations of disrespect.

Moreira (2015) mentions that:

> Although it is not directly related to the elaborate activities, something worth highlighting occurred at the end of the class. Two students disagreed, and the intervention was necessary so that no physical conflict occurred. This situation demonstrates the difficulty in keeping track of all the parameters involved in implementing a didactic sequence, in which unexpected attitudes end up happening (MOREIRA, 2015, p.59).

However, overcoming these obstacles is essential in the teaching-learning process. Running away from purely "content" classes (board and brush) where the mechanization of teaching and learning is still common in many municipal public schools is an obstacle.

Figure 12: Students talking about molecular models.

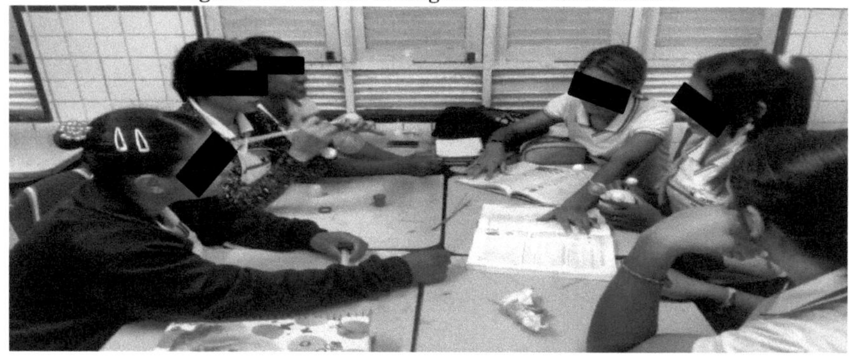

Source: Researcher data collection.

In professional and social life, as well as in the family, it is necessary to perform certain actions that we need to know how to express ourselves, consult, question, make plans, make decisions, make commitments and share tasks. These actions, involving practical, ethical and aesthetic aspects, may be relatively simple, other times they are complex. In the school space, group activities would qualify for challenges like these, so necessary in social life. This is why the importance of this activity is important where students participate actively and interactively in overcoming these obstacles.

FRIDAY (6TH) CLASS

In this class we had the purpose of using the models made by the students in an activity to complement the teaching-learning process aimed at using the didactic sequence. Using the models made we elaborated a final activity where the students described with their words which contents the models would explain and which would not be possible.

From this activity we have selected some answers from the students that we will use to continue our speech regarding their performance after using the didactic sequence as a teaching aid for the teaching of the content of Chemical Links.

Figure 13: Activity on the content of Chemical Links.

A partir dos modelos de Ligações Químicas confeccionados, responda o seguinte questionamento:
Quais aspectos os modelos confeccionados podem explicar e quais não podem explicar quando utilizados para estudar o conteúdo de Ligações Químicas?

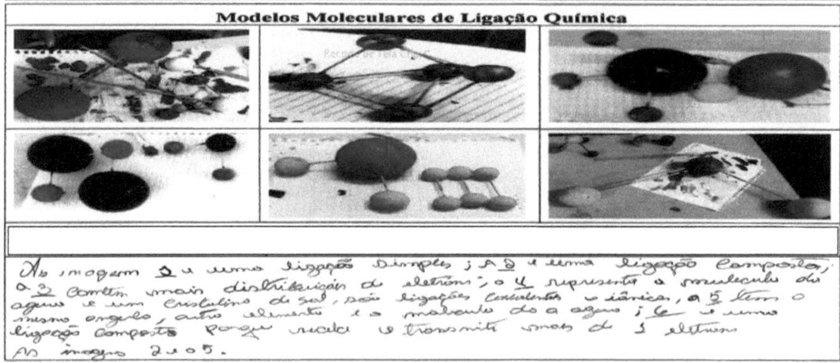

Source: Researcher data collection.

We can see in the "question 1" of the proposed activity, that the students from the manufactured models, make a correlation of some chemical contents, such as, electronic distribution, angle of formation of the molecules, molecular formula, etc.

They are able to make an association (possibly students listed the images from 1 to 6) of the models to the formation of single and compound links (which we can interpret as forming double or triple links).

In the same activity (question 1), we also asked students to describe the limitation(s) of the models developed. The answers of some students are well argued in their placements, for example:

> *"The aspects to explain the angle shapes" and the "toothpicks to differentiate the distance of each one", not to explain with the distances of the toothpick (T/U students).*

We found that students perceive the limitation of the made models, since it would not be possible to accurately determine, for example, the measurement of the bond angle of the atoms. Concepts such as molecular size, the radius of formation of molecules and molecular geometry (structure), could not be expressed precisely, we showed from the transcribed passage *"not to be explained with the distances of the stick",* where the students understand this limitation.

In order to evaluate the students' understanding of the use of the materials for making the models, we asked in "question 2" of the proposed activity that they describe in rich detail.

Figure 14: Activity on the content of Chemical Links.

2) A partir dos modelos de ligação química confeccionados, explique (descrevam com riqueza de detalhes) o uso dos materiais (bola de isopor, cores, palitos, etc.) utilizados e a forma estrutural dos modelos.

Source: Researcher data collection.

We see that the students perceive the representative functionality of the materials used in the representation of the elaborate molecular models, where Styrofoam balls are used to represent the atoms; the sticks representing the Chemical Bonds and the structural shapes of the molecules elaborated in the activity, for example, the water molecule mentioned by some students; and the use of colours for differentiation of atoms. How can we follow up on the transcribed answers of some students:

> *They serve to represent the chemical bonds and it is better to understand how the little balls together show that they are sharing electrons and colour to differentiate (Students X/Z).*

> *In image 4, small balls represent negative electrons and large balls represent positive electrons. And the colours, black with red represent an ionic bond, and black with green a covalent bond. One is crystalline with salt which is 2 large balls and 2 small balls and the other is a molecule of water (U/F students).*

> *The sticks were used to bind the isopo balls to each other the little balls were used to represent the ties of each bond and each colour to distinguish (A/D).*

Some important points deserve to be discussed further. For example, the fact that students understand that colours are used to differentiate atoms can be used as a starting point for the study of more complex chemical contents, such as Electronic Excitation (absorption and emission of energy). This subject can be contextualized and highlighted in the students' daily life when they witness, for example, the burning of fires at June parties or at the end of the year.

Another relevant aspect is the response of the *"U/F Students"* who can relate the models created with the ionic and covalent bonds, where they make reference to a crystalline salt reticle, but they cannot understand the representative functionality of the materials used, when they state that the smaller Styrofoam balls represent the negative electrons and the larger balls represent the positive electrons, when they were used to represent the atoms. In this sense, we must be careful not to create what Bachelard (1996) calls epistemological obstacles.

59

Finally, in "question 3" we asked students to summarise their learning about the content of Chemical Links, where they would describe the characteristics, elements and examples of the three types of links addressed. From the production of the students, we present some summaries below, so that we have a more efficient view of the learning presented by the students with the Didactic Sequence intervention, Table 2 below presents some summaries:

Table 2: Synthesis of learning about the content of Chemical Links.

TYPE	CHARACTERISTICS	ELEMENTS	EXAMPLES	PUPILS
Iônica	Element that donates and the other receives;	Non Metals + Non Metals	NaCl	D/J
	They donate electrons to stay stable;	Na: earthy alkaline metal; Cl: Halogen;	Sodium chloride	O/P
Covalente	In this bonding a sharing of electrons occurs to acquire stability;	Metals + Non Metals	H2O	D/J
	They are those that produce molecules.	Hydrogen; Oxygen.	H2O	O/P
Metallic	It is a bond that only occurs between metals;	Metals + Metals	Fe	D/J
	They are metals;	Iron	Fe	O/P

Source: Researcher data collection.

Based on the synthesis, we see that the students manage to conceptualize some characteristics, even if superficial, of the connections, pointing out some elements, or even families of elements, such as the "*alkaline metals*" which are linked to other chemical elements, for example, the "*halogens*" can form the ionic connections.

However, these do not mention other important characteristics, such as the formation of the crystalline reticulum in the ionic bonds, or even, the metallic alloys coming from the metallic bonds, or even, they cite the structures of resonance (subject addressed in the textbook: Being a Protagonist - Chemistry 1, used by teachers in that school).

The limitations (difficulties in forming a concrete concept) highlighted in the students' answers to this "question 3" and observed in some parts of the previous questions of this "final activity" to conceptualize the chemical processes that occur in the formation of chemical bonds, are also experienced and described by Batista *et al* (2013):

> We can understand the great difficulty of the students in identifying the processes that happen... Although the content has been worked on... it was possible to observe that this content still presents a high degree of complexity, besides the students' need to bring as background basic knowledge of

chemistry from the previous series to understand the content in question. Another criterion that influenced these results was time, as the didactic sequence would need more time to be worked on in order to obtain more significant results (BATISTA et al, 2013, p. 1-12).

However, the use of the Teaching Sequence to teach the content of Chemical Links, as well as others, results in a valuable tool for the teaching-learning process.

Batista *et al* (2013), concludes his research by saying:

> The proposal of this teaching methodology was a valuable resource, capable of promoting greater participation of the students in the work with the teaching of chemistry, collaborating for the motivation and stimulation of the students in the school. It is noted that students, even though they presented many learning difficulties in the content studied, still felt motivated and participated actively in the activities planned in the sequence, which contributed to raising their self-esteem from the moment that the content taught made sense by helping them to understand the role of chemistry and solve their everyday problems (BATISTA et al, 2013, p. 1-12).

It is with this vision that we conclude our analysis of the results, aiming at a more effective, meaningful and contextualized learning for the Teaching of Chemistry.

6. CONCLUSIONS

After analysing the results we can conclude that the objectives have been achieved, as the students are able to critically position themselves and to conceptualize with regard to the content of Chemical Bonds. It was possible to observe the involvement of the students during the realization of the Didactic Sequence steps. Their active and interactive participation in the activities that were developed. The students presented a satisfactory performance during the didactic path followed. The difficulties evidenced in the analysis do not make this strategy less viable for a teaching that allows a more efficient learning.

A relevant aspect experienced during the period of research in the classroom, and often common to the teaching exercise, is the short and interrupted time for the development of the Didactic Sequence, having at certain times faced some paralysis due to commemorative dates, national and state strike, period for applying tests, among others, we can see that the teaching methodology used consists of a very important supporting tool in the process of schooling the subjects targeted by the research.

Another obstacle encountered in the development of the Didactic Sequence was the time at which we carried out the proposed activities. It is recurrent in the period of the last classes (5th and 6th classes) the students' evasion. However, during the stages of the didactic sequence we noticed that the majority of the students remained in the classroom. As we have seen before two conditions are fundamental for effective learning, the use of significant material, and the willingness of the individual to learn.

From this perspective, we believe that it is the students who give meaning to the methodology used by the teacher (however it is not any method, didactic resource, or in any way that should be used), since the motivation is internal, what we seek is to arouse the interest of the student, his investigative capacity and criticism about the world around him.

Thus, we hope that didactic proposals of this nature can be used in the teaching-learning process with the objective of allowing a reflective, contextualized, interactive Chemistry Teaching, aiming at the personal, professional and social formation of the student so that he can exercise his citizenship.

REFERENCES

ALMEIDA, P. C. A.; BIAJONE, J. **Saberes professores e formação inicial de professores: implicações e desafios para as propostas de formação.** Education and Research, São Paulo, v.33, n.2, pp.281-295, May/August. 2007.

ANDRÉ, M.E.D.A. **Ethnography of school practice.** Campinas - SP: Papirus, 1995.

ANDRÉ, M.E.D.A. Research in everyday school life. In: FAZENDA, I. (org.). **Methodology of educational research.** 4 ed., São Paulo: Cortez, 1997.

AUSUBEL, D. P. **Educational psychology: a cognitive view.** Google translator, translated in 2017. Translation of the original the **Educational psychology: a cognitive view.** New York, Holt, Rinehart and Winston. 1968. p. 79.

AUSUBEL, D.P. (2003). **Acquisition and retention of knowledge.** Lisbon: Plátano Edições Técnicas. Translation of the original the **acquisition and retention of knowledge** (2000).

AYRES, C; ARROYO, A. **Application of a didactic sequence for the study of intermolecular forces using computer simulation.** Experiences in Science Teaching, V.10, No. 2, 2015.

ARAUJO, D.L. **What is (and how do you do) a didactic sequence?** Entrepalavras, Fortaleza - year 3, v.3, n.1, p. 322-334, jan/jul 2013.

BACHELARD, G. **The formation of the scientific spirit: contribution to a psychoanalysis of knowledge.** Trad. Estela dos Santos Abreu. Rio de Janeiro: Counterpoint, 1996.

BATISTA, A. *et al.* **Elaboration and evaluation of a teaching sequence for the content of electrochemistry.** In: Proceedings of the III Meeting of Initiation to teaching of UEPB, 2013. Pages 1 to 12.

BAPTISTA, J. A. *et al.* **Training of Chemistry Teachers at the University of Brasilia: Construction of a Curricular Innovation Proposal.** Química Nova na Escola, v.31, n.2, pp.140-149, May 2009.

BRAZIL. **National Education Guidelines and Bases Law**, Law No. 9.394, of 20 December 1996.

______ . Ministry of Education (MEC), Secretariat of Media and Technological Education (Semtec). **National Curricular Parameters for Secondary Education.** Brasília: MEC/Semtec, 1998.

______ . Ministry of Education (MEC), Secretariat of Media and Technological Education (Semtec). **National Curricular Parameters for Secondary Education.** Brasília: MEC/Semtec, 1999.

______ . CNE/CP Opinion No 9 of 8 May 2001 . **National Curricular Guidelines for the Training of Teachers of Basic Education, at higher education level, degree course, full degree.** Brasília. 2001.

______ . Ministry of Education (MEC), Secretariat of Media and Technological Education (Semtec). **PCN + High School: educational guidelines complementary to the National Curricular Parameters - Nature Sciences, Mathematics and their Technologies**. Brasília: MEC/Semtec, 2002.

______ . **RESOLUTION NO 2 OF 1 JULY 2015. CNE/CP resolution 2/2015**. Official Gazette, Brasilia, 2 July 2015 - Section 1 - pp. 8-12.

CHACON, E. P. *et al*. **Chemistry in the kitchen: possibilities of the subject in initial and continuing teacher training**. Revista Brasileira de Ciência e Tecnologia (R. B. E. C. T.) vol. 8, no. 1, jan-abr. 2015.   ISSN - 1982-873X

CHASSOT, A. **Catalysing transformations in education** - Ijuí: Ed. UNIJUÍ, 1993, p. 37 - 112.

. **Scientific Literacy**. Ijuí: Ed. Unijuí, 2001.

______ . **On probable models of atoms**. New Chemistry at School, n. 3, May 1996.

COSTA, M. C. **Educomunicar é preciso**. Nucleus of Education and Communication of the University   of   São   Paulo.   2006.   Available   at: http://www.usp.br/nce/aeducomunicacao/saibamais/textos. Accessed on 24/11/2016.

DOLZ, J.; NOVERRAZ, M.; SCHNEUWLY, B. **Didactic sequences for oral and written: presentation of a procedure**. IN.: SCHNEUWLY, B.; DOLZ, J. **Oral and written genres at school**. [Translation and organization Roxane Rojo and Glais Sales Cordeiro] Campinas, SP: Mercado de Letras, 2004, p. 95 - 128.

FERNANDEZ, C; and MARCONDES, M. E; **Students' Conceptions on Chemical Bonding**. New Chemistry at School, v. 24, p.20-24, 2006.

FIRME, R. N.; RIBEIRO, E. M.; BARBOSA, R. M. N. **Analysis of a didactic sequence on batteries: a CTS approach in the chemistry classroom**. XIV National Meeting of Chemistry Teaching. Curitiba-PR: [n.°]. 2008.

GARCIA F.A.; and GARRITZ, R.A.; **Development of a teaching unit: The study of the chemical connection in high school**. Teaching of science. Google translator, translated in 2017. Translation of the Original El **Desarrollo de una unidad didáctica: El estúdio Del enlace químico en bachillerato**. Enseñanza de las Ciencias, v. 24 (1), p. 111-124, 2006.

GOMES, H. J. P.; OLIVEIRA, O. B. **Epistemological obstacles in science teaching: a study of their influences on atom conceptions**. Science & Cognition, Vol.12, p. 96 - 109, Nov., 2007.

KOBASHIGAWA, A.H.; ATHAYDE, B.A.C.; MATOS, K.F. de OLIVEIRA; CAMELO, M.H.; FALCONI, S. **Estação ciência: formação de educadores para o ensino de ciências nas séries iniciais do Ensino Fundamental**. In: IV National Seminar ABC in Science Education.   São   Paulo,   2008.   p.   212-217.   Available   at: <http://www. cienciamao.usp.br/dados/smm/_scienciaformacaca   de   educadores   para

aensindeciêncianasseriesin the initial series of basic education. trabalho.pdf>. Access on: 05 Oct. 2016.

LANDEIRA, J.L. **Didactic Sequence**. Available at: https://www.youtube.com/watch?v=aqyXc7KIkDs. Access on: 20 March 2017.

LIMA, A. A; NUÑEZ, I.B. Learning by Models: Using Models and Analogies. In: RAMALHO, B. L.; NUNEZ, I. B. **Fundamentals of Teaching Learning Natural Sciences and Mathematics: The New High School,** 1st ed. Sulina, 2004, p. 245-264.

LINCOLN, Y. S.; GUBA, E. G. *Naturalistic inquiry*. Newbury Park: Sage, 1985.

LINS, R. C.; GIMENEZ, J. **Perspectives of arithmetic and algebra for the 21st century.** Campinas: Papirus, 2001.

Lisbon. J.C.F (organiser). **Chemistry, 1st year: high school.** 1. ed. - São Paulo: SM Edition, 2010. (Ser Protagonista Collection)

LOPES, A. R. C. **Textbook: obstacle to learning chemical science**. New Chemistry, V. 15, N. 3, p. 254 - 261, sea, 1992.

LOPES, A. C.; **Curriculum and Epistemology**; Editora Unijuí; 2007.

LUDKE, M.; ANDRÉ, M. **Research in education**: qualitative approaches. São Paulo: Editora Pedagógica Universitária - EPU, 1986.

MACHADO, A. R.; CRISTOVÃO, V. L. A **construção de modelos didáticos de género: contribuições e questionamentos para o ensino de género**. Language in (Dis) course, volume 6, number 3. set/dez., 2006.

MAIA, P. F. *et al*. **Modeling and representations in the teaching of ionic bonds: analysis in a teaching strategy**. Annals of the VI ENPEC, 2007.

MÉHEUT, M.; PSILLOS, D. **Sequences of teaching-learning: objectives and tools for research in scientific education**. Google translator, translated in 2017.
Tradução do original the **Teaching-learning sequences: aims and tools for science education research**. International Journal of Science Education, 26, n. 5, 2004. 515-535.

MELO, M. R; NETO, E. G. L. **Teaching and Learning Difficulties of Atomic Models** Vol. 35, No. 2, p. 112-122, May, 2013. New Chemical Magazine at School.

MORALS, C. M. V.. **Multimedia resource "Moleculito": Example of construction and evaluation in Basic Education**. Porto, 2007.

MORAN, J.M. **The video in the classroom. Communication and education.** São Paulo, v.1, n.2, p. 27-35, Jan./abr. , 1995.

MOREIRA, R. H. **Proposed didactic sequence using diverse resources for teaching and learning specific astronomy topics.** Dissertation (Masters) - Universidade Federal de São Carlos, 2015. 186 f.

MORTIMER, E. F. **Language and concept training in science teaching.** Belo Horizonte: UFMG, 2000.

MORTIMER, E. F. **The meaning of chemical formulas**. Revista Química Nova na Escola, no. 03, May 1996, p. 19-21. Evolution of atomism in the classroom: changing conceptual profiles. São Paulo, 1994.

MANTOVANI, S. R. **Didactic sequence as an instrument for significant learning of the photoelectric effect.** Dissertation (master) - Universidade Estadual Paulista, Faculdade de Ciências e Tecnologia. Presidente Prudente: [n.º], 2015.

MOURA, M. O.; MORETTI, V. D. **Investigating the learning of the concept of function from previous knowledge and social interactions**. Science & Education, v. 9, n. 1, p. 67-82, 2003.

NASCIMENTO, F.; FERNANDES, H. L.; MENDONÇA, V. M. **Teaching science in Brazil: history, teacher training and current challenges**. HISTEDBR On-line Magazine, Campinas, n.39, p. 225-249, Sep. 2010 - ISSN: 1676-2584.

PARIZ, E. **Metal connection: a proposal for teaching aids for the classroom teacher**. Dissertation (master's degree) - University of Brasilia, Postgraduate Program in Chemistry Teaching, 2011.

PEREIRA, J. E. D.; ALLAIN, L. R. **Considerations about the researcher teacher: to what research and to which teacher does this training proposal refer?** Teacher's view, Ponta Grossa, v. 9, n. 2, p. 269-282, 2006.
POLIDORO, L. F.; STIGAR, R. **A Transposição Didática: a passagem do saber científica para o saber escolar**. Ciberteologia - Revista de Teologia & Cultura - n. 27, 2006.

RAMALHO, B. L.; NUÑEZ, I. B.; GAUTHIER, C. **Training the teacher, professionalising teaching: professionalising teaching prospects and challenges**. Porto Alegre: Sulina, 2004.

REIS, D. B. **Diagnosis of the difficulties presented by Chemistry teachers for the use of models in public schools in the city of Campina Grande-PB**. Monograph (undergraduate) - University of Campina Grande: UEPB, Graduate Program in Chemistry Teaching, 2013.

RICHARDSON, R. J. **Social research: methods and techniques.** São Paulo: Atlas, 1999.

SANTOS, A. C. S. **Complexity and training of chemistry teachers. In: Brazilian Meeting of Complexity Studies.** Curitiba, 2005. Annals... Curitiba: PUCPR, 2005.

SANTOS, A. S. *et al.* **The public image of chemistry presented in the online articles of Ciência Hoje magazine.** Ex@tas Online, Vol. 6, n. 1, pp. 49-67. ISSN 2178-0471. Apr. 2015.

SILVA. C. S.; L. OLIVEIRA, L. A. A. **Initial training for chemistry teachers: specific and pedagogical training. IN: NARDI, R. (org.) Teaching of science and mathematics, I: subjects on teacher training [online].** São Paulo: UNESP Publishing House; São Paulo: Cultura Acadêmica, 2009. 258 p. ISBN 978-85-7983-004-4. Available from SciELO Books http://books.scielo.org

SILVA, L. L. and TERRAZZAN, E. A. **Correspondences established and differences identified in didactic activities based on analogies for the teaching of atomic models.** UFSM, 2008.

SOUZA, V. C. A.; JUSTI, R. and FERREIRA, P. F. M. **Analogies used in teaching the atomic models of Thomson and Bohr: a critical analysis of what students think from them.** Rev. Investigations in Science Teaching, v. 2, n. 1, p. 7-28, 2006.

VARGAS, S. L.; MAGALHÃES, L. M. **The genre tirinhas: a proposal of didactic sequence.** Educ. foco, Juiz de Fora, v. 16, n. 1, p. 119-143, mar. / aug. , 2011.

ZABALA, A. **Educational practice: how to teach.** Trad. Ernani F. Da F. Rosa. Porto Alegre: Artmed, 1998. p. 53-87.

ZUNINO, A. V. **The teacher-researcher interfaces and the teaching and learning process of natural sciences.** Research acts in education - PPGE/ME FURB, v.1, n.1, p. 53-74, 2006.

d) Students can use the text produced, consult their notes as well as the textbook and other sources of information.

e) Regroup the students in a circle to display and explain the models they have made.

f) Ask students what they think of the model they have created, if it is a good analogy? Why is it a good analogy? What aspects of your model do you explain? What aspects of your model do you not notice explain?

- **Activity 3: Evaluation**

a) We tried to verify whether the students acted as active subjects in class. We observed their involvement in the proposed activities and whether they performed what was proposed.

b) The evaluation process was continuous. However, at the end of the application of the didactic sequence we proposed an individual activity to the students, complementing the continuous evaluation.

APPENDIX III

We describe below how the classes were planned and developed, pointing out the objective(s) with each activity carried out in the steps followed by the didactic sequence:

AULA 1

ACTIVITY 1: Presentation of the Didactic Sequence.

TIME: 10 minutes.

COMMUNICATION APPROACH: Dialog and Exhibition.

PURPOSE: Since it is necessary for the activities to be ordered, structured and articulated in such a way that the stages, as well as the objective(s) are known both by the teacher and by the students, we have made the presentation of the Didactic Sequence.

SUPPORT MATERIALS: Power Point.

DESCRIPTION: At the 1st (first) moment with the students we explain the didactic sequence that would be worked on.

ACTIVITY 2: Application of the Objective Questionnaire.

TIME: 30 minutes.

COMMUNICATION APPROACH: Practice.

PURPOSE: An important point discussed in significant learning is the previous knowledge of the student. From the application of these questions we aim to know in advance the students' knowledge about the content.

SUPPORT MATERIALS: Sheet of Paper and Pen.

DESCRIPTION: Application of the written activity (questionnaire) composed of questions to mark. Division of the class into pairs of students.

AULA 2

ACTIVITY 1: Discovering Chemical Connections.

TIME: 40 minutes.

COMMUNICATION APPROACH: Dialog and Exhibition.

PURPOSE: Our aim was to hold an informative and dynamic class, where students could interact in an active way, through the discourses (questions) that arose around the subject addressed.

ACTIVITY ON THE CONTENT OF CHEMICAL LINKS

ESCOLA ESTADUAL DE ENSINO FUNDAMENTAL E MÉDIO MONSENHOR JOSÉ BORGES DE CARVALHO.

Alunos(as): _______________________________________ Turma: _________

partir dos modelos de Ligações Químicas confeccionados, responda o seguinte questionamento:
Quais aspectos os modelos confeccionados podem explicar e quais não podem explicar quando utilizados para estudar conteúdo de Ligações Químicas?

Modelos Moleculares de Ligação Química

Ligações Químicas? Que aspectos os modelos confeccionados não dá conta de explicar?

2) A partir dos modelos de ligação química confeccionados, explique (descrevam com riqueza de detalhes) o uso dos materiais (bola de isopor, cores, palitos, etc.) utilizados e a forma estrutural dos modelos.

3) A partir da tabela abaixo sintetize seu aprendizado sobre o Conteúdo de Ligações Químicas.

Tipo	Característica	Elementos	Exemplo
Iônica			
Covalente			
Metálica			

Na natureza nada se cria, nada se perde, tudo se transforma.
(Antoine Lavoisier)

Printed by Books on Demand GmbH, Norderstedt / Germany